Praise for

EVERY TOOL'S A HAMMER

"The handyman dad will *love* this book. . . . Adam Savage challenges readers to take a deeper look at what inspires them when it comes to 'making and molding, building and breaking' offering his own tips and tricks along the way about his favorite techniques and tools."

—*O, The Oprah Magazine*

"A personal look at the importance of creativity in all walks of life."

—*The Verge*

"Adam has drawn for us an imperative how-to for creativity that goes well beyond scissors, saws, and glue, to include vulnerability, self-confidence, and self-deprecating humor. I am aware of no human outside of fiction more qualified to pen this rousing paean to making. I adore this book."

—Nick Offerman, *New York Times* bestselling author of *Paddle Your Own Canoe*

"This book is creative rocket fuel. Adam is a master maker, and this might be his greatest creation yet—a funny, vulnerable, and soulful dive into the beautiful mind of a passionate artist. It's about making stuff, but there is also philosophy, insight, and, most of all, inspiration."

—Ed Helms

"Adam has stocked up a lot of deep thought and deeper wisdom: about how to make things large and small, how to make decisions, and how to make sure you're making the things that matter. Consider this book as a 3-D printout of Adam's brain, and be glad you have it."

—John Hodgman, *New York Times* bestselling author of *The Areas of My Expertise* and *Vacationland*

"Artists, inventors, and creators of every stripe will find Savage's work inspiring and informative, while *MythBusters* fans and others will savor his many amusing 'making' foibles and misadventures."

—*Booklist*

EVERY TOOL'S A HAMMER

Life Is What You Make It

ADAM SAVAGE

ATRIA PAPERBACK

New York • London • Toronto • Sydney • New Delhi

An Imprint of Simon & Schuster, Inc.
1230 Avenue of the Americas
New York, NY 10020

First Atria Paperback edition May 2020

ATRIA PAPERBACK and colophon are trademarks of Simon & Schuster, Inc.

For information about special discounts for bulk purchases, please contact Simon & Schuster Special Sales at 1-866-506-1949 or business@simonandschuster.com.

The Simon & Schuster Speakers Bureau can bring authors to your live event. For more information or to book an event, contact the Simon & Schuster Speakers Bureau at 1-866-248-3049 or visit our website at www.simonspeakers.com.

Interior design by Timothy Shaner, NightandDayDesign.biz

Manufactured in the United States of America

5 7 9 10 8 6

Library of Congress Cataloging-in-Publication Data is available.

Photo Credits: Page 57: credit Prop Store of London. Pages 76, 109, 254, 255: © Industrial Light & Magic. Used with Permission. Page 124: credit Michael Shindler. Pages 159, 200: credit Norm Chan.

ISBN 978-1-9821-1347-6
ISBN 978-1-9821-1348-3 (pbk)
ISBN 978-1-9821-1349-0 (ebook)

For my family, and all the other makers of the world

CONTENTS

EVERY TOOL'S A HAMMER

INTRODUCTION

Making is more than the physical act of building. It's dancing, it's sewing. It's cooking. It's writing songs. It's silk-screening. It's breaking new trails both literally and figuratively. Making, as my friend Andrew Coy, the former White House Senior Advisor for Making* under President Obama, says, is simply a new name for one of the oldest human endeavors: creation.

Making things has driven me ever since I can remember. It's also been my employment forever, or nearly so. First, as a jack of all trades in the theater scene in New York and San Francisco in the mid-1980s and early 1990s; then as a model maker for commercials and movies; and finally for a solid fourteen-year stretch as a producer, science communicator, and serial blower-upper of things on *MythBusters*.

When people who've achieved some public success write about their lives, often their experiences get charted like a purposeful, linear climb up a mountain toward a summit of achievement. There is often the perception that all of that person's life was driven toward some goal either by fate or personal ambition. Whether it's winning an Olympic medal or founding a Fortune

* Yes, that was his actual title.

500 company or going to the moon, the story always seems to arc the same way. Life stories always look like straight lines from the vantage point of looking back, but precious few really are. My story certainly isn't.

My story is more of a path with many forks. There was a general direction I wanted to head, and a vague sense where I wanted to end up—LEGO designer! *Star Wars* special effects guy!—but at each of those forks, when I actually reached them, the decisions I made in the moment were based mostly on circumstances and opportunities that were directly in front of me. Some turns were wrong, some turns were right, some turns were just weird but then became right with time, like with *MythBusters*.

The fandom that Jamie Hyneman and I encountered from *MythBusters* was by no means broad-based, but as narrow as it was, it was twice as deep, because at that time there just wasn't a robust making community for young creatives to tap into. This is not to say that I was some kind of trailblazer. In fact, quite the opposite. I was following well-trod paths carved by generations of makers who came before me. But I think one of the reasons the show became so popular is that what we were doing was kind of an anomaly. Even though we could tell from the fans that there was still a love of making out there in the ether, it seemed that fewer people were taking up hammers to do the kinds of things that I was interested in. There was a lack of young people who were getting any practice working with their hands and their hearts to make things that were important to them. To create.

There are probably a million factors that contributed to why this was happening—the rampant elimination of high school and early-grade shop classes through the 1980s and 1990s, an over-fixation on graduate degrees, a focus on technology and/or finance as primary modes for upward mobility, too many screens.

I'm not a sociologist or an anthropologist, so I don't have a full explanation for what I was seeing, just that I was finding it more and more difficult to find meaningful populations of good young makers to share ideas with.

Sometime in the mid-2000s that began to change, thanks in part to advances in rapid-prototyping technologies like 3-D printers, open-source software, and the spread of broadband internet. This DIY *maker* movement that emerged empowered young people, underprivileged communities, and the simply curious to learn, and teach, and share how to make things again. I also deeply credit Dale Dougherty, who founded *Make:* magazine in 2005 and offered a vision of an updated *Popular Mechanics* that felt like it was pulled directly from my wildest dreams. It was the perfect flagship for making, celebrating a broad constellation of creativity defined by tackleable projects and learnable practices.

Maker Faire was founded soon after in San Mateo, California, and a community was born. I'm proud to say I've been a part of the faire from the very beginning and I've given a talk almost every year since its inception. Over time it's come to be known as my annual Sunday sermon (a name I was unaware of for the first several years). Every year my subject is different, but inevitably I conclude with some kind of exhortation to keep making stuff, to keep creating, and to keep pushing past self-perceived limits. Because more than anything else, what I continue to fight against is all the ways in which the tools of creation are kept out of the hands of our most dynamic, creative minds. Whether it is because of apathy, lack of access, bureaucratic inertia, community indifference, or educational redlining, I don't care. The world needs more makers.

After the talk, I take a couple hours to meet fellow makers. It's always my favorite meet and greet of the year. We share stories and take pictures and I ask people what they're making, because even if they're nervous, their enthusiasm for what they're currently building always wins out. Give a maker the chance to tell you about the thing they're putting their time into, and good luck getting them to stop!

In one of the first years of the Maker Faire, a young man came up and said kind of sadly, "I don't make, I code." I've heard this sentiment a lot. "I don't make, I __________." Fill in the blank. Code, cook, craft (not sure why all my examples start with C), the list of exceptions people invent to place themselves on the outside of the club of makers is long and, to me, totally infuriating. Because the people who do that to themselves—or more likely, the people who TELL them that—are flat wrong.

"CODING IS MAKING!" I said enthusiastically to that young man. Whenever we're driven to reach out and create something from nothing, whether it's something physical like a chair, or more temporal and ethereal, like a poem, we're contributing something of ourselves to the world. We're taking our experience and filtering it through our words or our hands, or our voices or our bodies, and we're putting something in the culture that didn't exist before. In fact, we're not putting what we make into the culture, what we make IS the culture. Putting something in the world that didn't exist before is the broadest definition of making, which means all of us can be makers. Creators.

Everyone has something valuable to contribute. It is that simple. It is not, however, that easy. For, as the things we make give us power and insight, at the same time they also render us vulnerable. Our obsessions can teach us about who we are, and who we want to be, but they can also expose us. They can expose our weirdness and our insecurities, our ignorances and our deficiencies. Even now, at fifty-one years old as I write this, I am enduring a frightening vulnerability with this book.

I didn't know how to write a book before I wrote this one. Like most of the things I've learned throughout my life, I learned about the writing process by going through it. It was surprisingly complex and fairly difficult. As I'll talk about in the pages to come, I like projects that have high levels of complexity, so in theory I should have been able to handle something like this, but the reality was that I was wholly unprepared for the Gordian knot inherent in organizing my thoughts over dozens of thousands of words. I count many friends who are published writers, whose livings come from making books, and my hat is off to them, because books are *hard.* They are scary. What is here is both deeply personal and, I hope, educational. And to be perfectly honest, I'm satisfied with the results of my efforts on a level I didn't expect. This is the risk of all creative spirits: every project has as many obstacles as solutions, and with each one there is the chance that one might not end up satisfied with the results; or that others might not be satisfied with them, either, and can't wait to tell you about it.

This is one of the main reasons I believe that adolescence can be so fraught for so many. Just as we start to catch the barest glimpses of our true selves and begin to understand what it is about the world that fascinates and intrigues us, we often run right into people who aren't ready to be encouraging and can be downright hostile to someone being "different." It can be a terrible early lesson in what is safe to share about ourselves. In this sense, proclaiming and revealing a deep curiosity—an obsession—to others is to show them our bellies. When my dogs roll on their backs and let me scratch their bellies, they're paying me the high compliment of their own vulnerability. They're showing me the deepest trust.

Kids, on the other hand, can be cruel. Not all of them, but enough that adolescence is when a lot of people learn to hide their true selves, to bury their creative interests and creative instincts, in an act of self-preservation. The tonic to this is to find people you can trust and show your belly to: a best friend, a social group, a *sangha*. In that regard, it's never been a better time in history than right now for finding your people. The internet is far from fulfilling its promise to be the compendium of all human knowledge, it's more like the outline or the index. But where it shines is in giving people all over the planet the ability to find a peer group of enthusiasts with which to share their creativity, and thus themselves. That's a net good. When we find our people, we find in them the permission to explore, to exult, and to share.

This book is my attempt to share my explorations with you. It is a chronicle of my life and the lessons I've learned along the way, and it is also a permission slip. The permission slip is from me, to you. It says you have the permission to grab hold of the things you're interested in, that fascinate you, and to dive deeper into them to see where they lead you. You might not need that permission. If that's the case, good for you! Go forth and do awesome things. But I have needed that permission many times in my life. And whenever I found it, it helped me uncover secrets about myself and about the world I live in. It made me better as a man, as a maker, and as a human being.

We are built to collaborate. Humans are explorers, and social creatures. We are driven to share our stories. Our stories are what make us so unique on this planet. I mean ostensibly unique. There might be great speculative fiction being promulgated among the octopus and cuttlefish communities, or delightfully wry oral histories being shared between orca, or gray wolves, but until we can decipher them it's humans alone who expand our understanding of the universe by swapping stories about what we see and have seen. Making is one of the principal ways we share, and have always shared, our stories.

The structure of *Every Tool's a Hammer* changed several times in the course of its writing. It's quite a different book than I thought it would be when I began, which is funny because that is, in fact, one of the throughlines of the book now that I can see it as a whole: that nothing we make ever turns out exactly as we imagined; that this is a feature not a bug; and that this is why we do any of it. The trip down any path of creation is not A to B. That would be so boring. Or even A to Z. That's too predictable. It's A to way beyond zebra. That's where the interesting stuff happens. The stuff that confounds our expectations. The stuff that changes us.

The book is broken into four main sections. The first deals with motivation and the physics of creativity. I consider healthy obsession to be the gravity that binds us to the things we make. I believe that to be excellent at anything you need to be at least somewhat obsessed with it, and this section explores how to use obsession to find ideas and execute them.

The second section deals with the notion of witnessing how you work, and noticing what it can tell you about how to work better. I look at how time can substitute for skill when engaged in the unfamiliar, and how small amounts of extra time invested in early stages can save massive amounts of time on the other end. Finally, I get into how to see yourself and your work habits in the context of your work and also how to widen that circle of awareness to include others so that you can share what you're doing with them. Making can often be solitary, but I find it's vastly more fun as a community exercise.

The third section deals with tolerance. Both as an engineering term and as a meta term of art. When we say we need to teach kids how to "fail," we aren't really telling the full truth. What we mean when we say that is simply that creation is iteration and that we need to give ourselves the room to try things that might not work in the pursuit of something that will. Wrong turns are part of every journey. They are, as Kurt Vonnegut was fond of saying, "dancing lessons from God," and the last thing we want to do is give our kids two left feet.

The fourth section has an organizational bent to it, specifically the organization of a maker's work space. I believe every shop is a physical manifestation of a maker's philosophy on how to work. And by understanding that philosophy, you can fine-tune your methods, your habits, and your builds.

In the end, *Every Tool's a Hammer* is an eclectic mix of story and instruction, which suits me. Eclecticism is kind of my brand. I'm a generalist in my creative output and a wannabe polymath in how I organize my life, so it's only fitting that for every cautionary tale and story of triumph, there is a lesson on the tools, techniques, and materials that have defined my life as a maker. Frankly, I thought there would be more of the latter, but the deeper I got into the writing, the more wary I became of speaking from a position of authority because my talent lies not in my mastery of individual skills, at which I'm almost universally mediocre, but rather in the combination of those skills into a toolbox of problem solving that serves me in every area of my life. It's important to mention that this toolbox includes a wide roster of incredible, inspiring makers and creators, whom I was lucky enough to consult during the writing of this book. Their honest and engaging discussion of their approach to their craft, kept inspiring me as I worked on this project and I hope it contributes toward inspiring you, too. Reading about making always makes my hands itch to make something myself. If this book has anything like that effect on you, I'll feel like I've done my job.

So let's get making.

DIG THROUGH THE BOTTOM OF THE RABBIT HOLE

"How do I get started?" Across four decades of making, I have been asked this one question more often than any other. It's a simple question on its face, with not so simple answers underneath. At an individual project level, my answer is usually "Well, it depends," in large part because creation and making have their own particular dynamics that involve unique concerns with the mental physics of inertia, momentum, structural cohesion, friction, and fracture. Thus, the rules of what you're making often determine how you begin.

Most of the time, however, the question really being asked is, "How do I get started when I have no idea what to make?" That's when the question moves from the physical world of making to the internal, mental space of ideation and inspiration. I have come to believe that the answer to this question resides within one of the grander, fundamental principles of physics, the first law of thermodynamics: an object at rest tends to stay at rest unless acted upon by an outside force. Which is to say, to get started *you* need to become the outside force that starts the (mental and physical) ball rolling, which overcomes the inertia of inaction and indecision, and begins the development of real creative momentum.

With my personal proclivity for speed and experimentation, I rarely have an issue getting moving, and rarely have difficulty coming up with ideas as a result. With eyes that have always been bigger than my stomach, my creative plate has been consistently full to overflowing with ideas. My battle is usually with time and resources more than worrying about what my next project will be.

I know this might make me unique in some maker circles, and probably infuriating to others, but I assure you that this has less to do with any special skill on my part and more to do with one specific trait: obsession. In my experience, bringing anything into the world requires at least a small helping of obsession. Obsession is the gravity of making. It moves things, it binds them together, and gives them structure. Passion (the good side of obsession) can create great things (like ideas), but if it becomes too singular a fixation (the bad side of obsession), it can be a destructive force. As a maker, which result you experience depends largely on how you discover, engage with, and manage the sources of your obsessions.

I am a serially curious person. Countless things have captured my attention over the years: history, science fiction, film, the architecture of public spaces, mechanical computers, glue, LEGOs, curse words, magic, storytelling, *Star Wars*, physics, philosophy, armor and weaponry, magic and monsters, new tools, tiny cars, space suits and spaceflight, animal consciousness, eggs. I've not found an end to the list of things that have sent me deep down various rabbit holes for exploration. Thankfully, I had early support from parents who cosigned many of these flights of fancy and encouraged my natural interests. My dad was an artist and my mom was a psychotherapist. I lucked out there. If I was curious about something, they gave me permission to explore it. When I didn't know how, they made the tools of exploration available to me. At one level, I think that what my parents were trying to do was to keep my curiosity aimed at something constructive, something other than mischief, though I was certainly able to engage in a fair amount of that in my time. In the house I grew up in, my folks put real value on following one's passionate interests wherever they might lead. They knew that if I would let those feelings be my guide, I would be more likely to *do something* with the fruits of that exploration.

Emotional self-awareness is a tall task for a kid. Hell, it's tough when you're an adult. It's hard to put words to emotions. It's even more difficult when verbalizing them in public might subject you to scorn. That was certainly the case for me. The pubescent teen me had no earthly idea how to describe what *Star Wars* or science fiction or the Apollo astronauts made me feel. At least not in a way that I wasn't sure would get me stuffed into a locker. So I kept my enthusiasms and feelings to myself. This is a strategy that is not unique to young, enthusiastic, creative types. Where I differed was that in keeping my feelings secret, I did not also bottle them up and extinguish them, as so often can be the case when you don't have a supportive environment at home. Instead, I simply let them multiply inside me until they were all that I could think about.

In this sense, what my parents had really done by nurturing my curiosity was to give the original green light to my creative obsessions, and I will be eternally grateful to them for that. Their encouragement demonstrated to me that my budding obsession was a thing of value, not a trifling thing to be dismissed; my fascinations were worth something; my curiosity was currency to be spent in the service of deep exploration, both of the external world and also of myself. They gave me license to pursue what I have called my "secret thrills."

FOLLOW YOUR SECRET THRILLS

Secret thrills can come from anywhere and anything at any time. If you happen to be a cinephile or an architecture fan like me, it might be the MacGuffin that pushes your favorite movie's plot forward, or it could be the verdigris patina of some weathered architectural detail on a building you pass every day on your way to work or to school. If you're paying attention, those types of things will catch your eye, and if you let them, they'll start to engage your mind. Once in a while they will even thrill you enough within the privacy of your own imagination to feed a desire to go deeper into that thing, to know more about it, possibly even to possess and do something with it. Budding (and matured) obsessions like these are where ideas come from.

In my experience, when you follow that secret thrill, ideas pop out from the woodwork and shake out of the trees as the gravity of your interest pulls you farther down the rabbit hole. And yet, so few of us give that thrill much purchase. We may even dismiss it as an indulgence or a distraction. There is almost a quiet shame in it, which is a big part of the reason why that secret thrill always seems to remain secret for so many. Over the years, I have lost count of the times people have come up to me and begun a conversation by quietly, almost reluctantly, admitting to their own curiosity about something I've done or a hobby I pursue. There is a belief among many of these types, that to jump with both feet into something like that is to play hooky from the tangible, important details of life. But I would argue—and have—that these pursuits *are* the important parts of life. They are so much more than hobbies. They are passions. They have purpose. And I have learned to pay genuine respect to putting our energy in places like that, places that can serve us, and give us joy.

I've been fortunate in that I've been able to follow my secret thrills into adulthood and then into professional success. But even if I hadn't been able to do that for a living, if I could only chase those thrills in my free time, I would still be constantly making stuff.

This stands in stark contrast to other fleeting interests and random skills that I used to pursue, like juggling or dramatic performance, which I gave up on once I got a notch better than mediocre. With so many of those early fascinations, I never knew how to push past that point of proficiency, and I didn't care enough to find out. I was the Patron Saint of Mediocrity+1.

When I realized in my early twenties that I could pursue, and maybe catch, real excellence at a high level of making, that is when I dove in headfirst. And that pursuit has radically improved my ability to incorporate the skills I already had with new skills I hoped to acquire. It's also made me more comfortable with acknowledging the limits, which are substantial, of what I can do. For instance, I would *love* to be a writer of screenplays. A screenwriter's way of seeing is a special thing. They have a unique type of brain, one that filters the world it experiences entirely through narrative and has, over time, become a highly tuned machine in the service of character construction, world building, and plot layering. Screenwriters are basically human 3-D printers for story.

But I've learned that is not how my brain works. I don't think in arcing, twisting plots. It's not that I wish it were different. I'm actually okay with how my brain works. I don't view it as a deficiency. I don't *need* to write screenplays. Each one of us ends up building different ways to interpret and recapitulate the world as we make our way through it. Each of us has a unique way we share our stories, which means that each of us comes to our ideas differently, and we each express them differently. This is the magic that makes culture.

How does your brain work? What is your secret thrill? How do you process your world? Screenwriting is simply one route to creating stories. The specific skill set my brain has equipped me with is one solidly within the realm of making physical *things*. It has served me very well, even if it doesn't end up yielding a screenplay. And I'm okay with that, because for me making stuff has always felt different. Making stuff utilized my brain like no other skill I'd learned. There was something special in the marriage between the structure of my brain and what I could do with my hands. When I made stuff, the world made sense to me. It felt like my superpower.

Foremost among my passions for making stuff has been cosplay. Cosplay is at its most basic the practice of dressing up as favorite characters from movies, fiction, and especially anime, but it's about so much more than just putting on a character's costume. Cosplay encompasses stepping *into* the character him- or her- or itself. I've come to understand it as more of a participatory community theater than a solo practice. I have a deep abiding passion for cosplay. It has been a constant source of thrills, and an endless fount of ideas for stuff to make as a result. Many of my favorite projects are a product of this interest. I am unabashed and unequivocal in my love for it. Now, at least. But it has not always been so simple. See, the thing about cosplay, or most any deep interest that produces these secret thrills, is that while it IS fun, it can also be complicated, because (and here might be a source for some of the secret shame around our enthusiasm) the things we love tend to make us quite vulnerable.

The seeds for my cosplay obsession germinated in high school—well before the word was even invented—when I started to fall in love with film as a form. The multisensory storytelling and layered world building blew my mind. This was the early '80s, an incredible time for a teenager interested in sci-fi adventures, space operas, and fantasy epics. They inspired me to create my own versions of costumes to bring their dream worlds closer to reality, to put myself into those narratives—in the privacy of my own home, of course. I would only let that secret joy out for public consumption on Halloween, when I had a built-in excuse for my creative inspiration. I suspect this is how it starts for a lot of people.

At sixteen, my dad and I built a full suit of armor inspired by John Boorman's film, *Excalibur*, and I wore it to school the day of Halloween. We spent weeks researching it and fabricating it from aluminum roof sheeting and what felt like a million pop rivets. I worked on it ceaselessly until it fit like a glove and I felt properly awesome in it. The only structural problem I encountered: I couldn't sit down. If I wanted to stay in costume AND see what the teacher was writing on the blackboard at the front of the classroom, I had to stand against the wall at the back of the room. It was a trade-off I was more than willing to make, and one that I was getting the better end of as far as I was concerned, right up until about halfway through third period, when I started to overheat, develop tunnel vision, and then slowly slide down the wall with a loud, deliberate scrape, until I passed out in a heap in the middle of a math lecture. It's more than a little embarrassing to wake up in the nurse's office covered in sweat, stripped to your underwear, wondering where your homemade armor went.

The following year I went a little lighter on the metal and made a piece of forearm armor as part of a costume inspired by *Mad Max 2: The Road Warrior*. I fashioned a vambrace from aluminum and added some cool labels and futuristic graffiti. Then I got it to look suitably, post-apocalyptically weathered by scraping it repeatedly on a dirty stone wall in my basement. On Halloween, I wore the whole rig to school with a beat-up leather motorcycle jacket and some heavy-duty *Mad Max*–like boots. It was what I'd later learn was called an "in-universe" costume—not canon, but within the canon—that felt as badass to wear as it looked. In fact, more so even than the full suit of armor, if that was possible.

My classmate Aaron begged to differ. He ribbed me about my costume, not ruthlessly, but enough to get my hackles up. Ordinarily when something like this would happen, my tendency toward conflict aversion would send me retreating into myself, into the space where my obsessions lived, but not this time. Wearing the costume made me feel powerful (as I would later discover cosplay often does), and imbued with the spirit of a character from a post-apocalyptic universe who had managed to survive through the end-times, I reared up and talked back. In my head—or should I say, in the head of the character I was inhabiting—that should have been the end of it. I had parried Aaron's thrust, then countered successfully with a thrust of my own. Aaron was vanquished.

Aaron disagreed. "Oh, check out Adam being all powerful with some metal on his arm!" he yelled derisively, much to the delight of our classmates.

With a single sentence, Aaron had pierced my armor. He saw through me so clearly, and he used it against me, exposing the part of me that I had mostly kept private. In that moment, I realized that this thing that had produced such a transformational feeling of empowerment could, if I let it, also be turned against me to make me feel just as vulnerable as it had made me feel strong. It was a lesson that I would relearn many times as I got older.

In 2009, for example, *MythBusters* set out to tackle a classic movie myth. Throughout film history, heroes and villains alike have made their escapes by leaping from the roofs and windows of tall buildings into the safety of dumpsters in alleys below them, then casually climbing out and running away. But how hard or soft are the contents of the average real-world dumpster? What is the ideal material to encounter when you make the actual jump? And if that ideal material is, in fact, in the dumpster, will it save your life? These were all questions we planned on answering.

When we plotted out the story, it became obvious that Jamie Hyneman and I would need to do the jumping ourselves. This led to one segment of the episode that involved training and a second that would include the actual experimentation. From a visual storytelling perspective, I wanted our outfits for each segment to be different. For the training sequence, the wardrobe team made us tracksuits with the words STUNT TRAINEE pasted on their backs with iron-on transfers. For the experimentation sequence, since I would be our official jumper, I thought a lot about what kind of wardrobe would look good on screen, but in a manner befitting the theme of the episode.

Sitting atop one of the structures at the Treasure Island Fire Training Facility in the middle of San Francisco Bay, where we were filming the episode, I stared out toward the East Bay and my gaze landed upon the now-defunct Alameda Naval Air Station. Alameda NAS was the shooting location for some of our biggest car-related mythbusting as well as a number of sequences in one of my favorite sci-fi franchises of all time: *The Matrix* with Keanu Reeves as Neo. That was it! Neo was an epic roof and window jumper. I could easily dress up as him, I thought, then make the twenty-foot jump from a rooftop into the dumpster, and it would look awesome. Neo's iconic long coat, chosen by the

Wachowskis for its cinematic qualities, would be equally cinematic for our show.

So I set about carefully assembling a reasonably accurate Neo costume without really telling anybody on the crew.

Long flowing coat: found on eBay.
Oakley Twenty XX sunglasses: check.
Knee-high motorcycle boots with lots of buckles: a quick trip to Haight Street in San Francisco checked that one off the list.

When it came time to shoot the experimentation sequence the following day, I ran over to my car to change. Pulling on each piece of the Neo costume induced a new thrill, but as I came around from behind my car into the full view of my crew, I could see many of them snickering, suppressing smiles. Here was that complicated moment, again. I was completely exposed. In those younger years, this would have been a slow-motion nightmare. My mind's eye would have translated the muted church giggles into open mockery, like Carrie at the prom. But there was no cruelty in the actions of the *MythBusters* crew. I had already worked with most of these people for half a decade, and we were family through and through. They were snickering because they could see so clearly just how *into it* I was.

Putting on the Neo costume exposed a deeply private part of myself to the crew that few of them had yet seen, a part about which I am always just a little bit embarrassed. But I quickly remembered why I was wearing the costume: I knew that the long *Matrix* coat would look *amazing* as it flowed behind me in the high-speed shot of me falling through the air into a dumpster full of foam. And boy howdy, it really did. But I also realized that I was having a conversation between two versions of myself: I was giving a shout-out to

This high-speed shot from MythBusters' *"Dumpster Diving" episode is still one of my favorites.*

high school me, telling him, "It's okay to let your freak flag fly"; and I was reminding grown-up me to keep moving toward this thing that I know is a little bit weird, and that I love for reasons I don't fully understand, because moving toward these things has been the engine of everything I've achieved in my life.

The Neo coat was the first elaborate costume I put together for the show, and it spurred countless other costumes for future *MythBusters* episodes, which in turn fueled ideas for future builds for Comic-Con and videos for Tested.com, my website devoted to the process and tools for making in all its myriad forms. What I was doing that day on Treasure Island, in a sense, was giving myself my own green light to follow the obsessions that had defined my youth—to follow the secret thrills they produced all the way to their ends, regardless of what I found there, because sometimes what you find there are the best ideas you'll ever have.

I am a maker by trade and a storyteller by temperament, but first I see myself as a "permission machine." In the beginning of his incredible essay "Self-Reliance," Ralph Waldo Emerson says: "To believe your own thought, to believe that what is true for you in your private heart is true for all men—that is genius." The essay and in particular that phrase hit me hard in the solar plexus when I first heard it at eighteen years old, and it continues to today. The deepest truths about your experience are universal truths that connect each of us to each other, and to the world around us. I have found this truth to be the key that unlocks those shackles of shame and self-doubt. It gives you the elbow room to fly your freak flag, the mental space to pay attention to the things that you're interested in. For the creator within all of us, this is the pathway to ideas and creation.

PARTICIPATE IN YOUR WORLD

Every single one of us is trying to make sense of the world—our place in it, and how everything fits together. We learn as much about ourselves and our surroundings from the stories we choose to tell, as from the stories others choose to tell us. I'll admit, sometimes the source of our own stories can be slightly embarrassing. I recognize cosplay is not exactly the most useful and selflessly noble endeavor in the world, and I don't delude myself into thinking that by doing it I'm necessarily making the world a better place. However, in paying attention to what engages me, and then sharing with others the process and product, what *I'm* doing is producing things that might spark ideas or inspiration in them, just as others' work can spark ideas for me. Paying heed to my secret thrills has been the thread that continues to weave together my journey as a maker. Engaging with what interests you shouldn't feel like crazy advice, but we all know it's not always the easiest path.

In addition to looking inside of yourself for ideas, I'm a firm believer in, and practitioner of, spontaneous inspiration. Arlo Guthrie, son of legendary folk singer Woody Guthrie and an incredible songwriter himself, once said that he doesn't believe that songwriters write songs. "Songs are like fish," he said. "You just gotta have your line in the water." If he just sits by a river and casts his line out into the flowing water, every now and then a song might swim by. And if he's lucky enough, or skillful enough, it will bite at his lure before it passes beyond his reach. Of course, it's never *that* simple, which even he recognizes. "It's a bad idea to fish downstream from Bob Dylan," Guthrie concluded. Dylan, somehow, has the best line, the best bait and hook, the best net, that Guthrie has ever imagined. It's a miracle anyone downstream of Bob Dylan has ever even *seen* a song.

This is a romantic notion for sure, but only to the extent that there is often a spontaneity and serendipity to creation that is humbling. Inspiration can arrive with a suddenness that makes it hard to take credit away from the universe for a seemingly random, weird, awesome thought that turns out to solve a problem that's been in front of you this entire time. Of course, there's usually a ton of advance work that goes into letting that kind of thought come into being: practicing the mechanics of one's craft, paying deep attention to the state of the art, working to confront and to solve ever more difficult problems, and being awake to the world you pass through every day. These are the behaviors common to experts in every field, behaviors that often materialize well before expertise is even a consideration.

I can remember the first time an idea came to me from thin air. I was five years old and I was supposed to be taking a nap (when all the best adventures happen). Instead, I snuck out of my room with my teddy bear Jingle—so named for a bell in his ear that roughhousing and planned obsolescence had turned into a click rather than a ring—and sneaked unnoticed out to my father's studio. My father built a wondrous (to the mind of a child) art studio in the garage behind our house in North Tarrytown, New York. It was both his place of business and his sanctum. He was a painter, animator, filmmaker, and illustrator, and the studio was filled to the rafters with books and paper and matte board and canvas—and tubes of paint of all sizes and varieties; boxes of charcoal, pastels, colored pencils, drafting pens; tools, and rulers, and everywhere photos and drawings that inspired him.

It held everything a curious, creative kid could possibly want, and my father had only a few rules for being in there. There was a general understanding that you weren't to mess around unsupervised, and there was one very specific prohibition that came with no ifs, ands, or buts: DO NOT TOUCH THE SINGLE-EDGED RAZOR BLADES. As a parent, the wisdom of that rule is self-evident. As a child, he may as well have said those blades were made of Halloween candy and Tooth Fairy money, because every time I walked into the studio, they called to me from their box at the back of his big worktable. Still, I kept my distance. I valued my access to this magical space more than whatever mythical powers resided within these little wafers of sharpened adamantium.

But I had a vision for Jingle that came to me while I was trying to fight off sleep (my five-year-old version of sitting by the river) that required using those blades. Jingle was already old, his right eye misshapen by an evening too close to the fireplace, his bell didn't work, his paw pads had worn off, and I wanted to create a picture of him the way he was when I first got him, so it would be like he was always brand new.

I pulled down a piece of construction paper from a shelf and laid it on my dad's table. I placed Jingle in the center of the paper, and carefully traced his outline. Then I filled in his facial features. I wanted people to know who it was so I wrote, as best as I could (which was not very well), "Jingle" down one leg and "Savage" down the other. The writing I achieved was the visual equivalent of talking with your mouth full. Next I tattooed pads on the bottom of his paws. I'd actually done this before on the bear himself, several times, but inevitably they always faded away with wear and tear. As I watched the black ink on my drawing dry into a dark paw pad that would never disappear, and I compared it to the last remnants of a pad still barely visible on Jingle's paw, I must have realized that I could actually make Paper Jingle into anything I wanted.

But what: Fireman Jingle? I don't know. Action Jackson Jingle with Karate Chop? Maybe.

Here's the thing about stuffed animal talismans that became obvious once I finally had kids of my own: they are objects that are separate from you, but they also *are* you. They are both sponge and mirror. They are a projection. The adventures you take them on narratively are *your* adventures. Onto them you project all the things you want for yourself, that you like about yourself, and what you *don't* like about yourself. They are your world.

It turns out, that is exactly how I treated Jingle, because what I added was a nifty blue vest, a snazzy belt with a gold buckle, and a Superman symbol on his chest for good measure. Why? Well, these were clearly the things that I wanted for myself. I wanted to be a superhero. I wanted to . . . I don't know . . . dress like Meathead from *All in the Family*? (In my defense it *was* the early '70s.)

With each embellishment, Paper Jingle became more real.

Paper Jingle Savage, circa 1972.

More of me. But he remained trapped in that rectangular piece of brown paper, he could only ever be a drawing. If I wanted him to be fully realized—and boy did I—I would need to free him from this paper prison. That is when the siren song of the single-edged razor blades finally reached my ears. All I had to do was pluck one of the blades from its box, peel off its protective cardboard wrapper, and carefully cut along the outline of Jingle's body, and he would no longer be simply a picture of a teddy bear, he would be a bear-shaped picture of a teddy bear—of *my* bear, Jingle Savage.

To my five-year-old mind, he would be real. So I gave into temptation, and Paper Jingle Savage was freed.

I ran inside to show him off to my dad, damn the consequences. I guess I was hoping that he'd think I used a pair of scissors, but he could tell immediately from the cuts at the neckline, where Jingle's head meets his body, that I'd used a razor blade to bring him into the three-dimensional world. As an adult, I must say I'm actually impressed with how well five-year-old me managed the contours I cut with that flat blade. At the time, though, I was mostly just surprised (and relieved) that my dad wasn't mad. In fact, he was happy enough with my creation that he framed it himself, which is why I still have it to this day.

I think the reason I escaped punishment was that, while I'd violated his primary rule about my using the studio, I'd actually held true to the true purpose of the space: I was in the studio for a reason, and I clearly wasn't messing around. I had an idea and I executed it. This was the first time that my youthful curiosity had led me to making instead of to mischief. That simple shift was all that my father was really waiting for. When it arrived, the rules relaxed, I gained more trust and access, and, unbeknownst to me at the time, a lifetime of creative exploration began.

I never have a clear sense for when inspiration like this will strike. It's often quite subtle, but I've become attuned to its presence because I've worked to remain connected to the world in which I live. This is one of the hardest things to do, for young makers especially. Trust me, I know, I've been there. When those interests that produce thrills within you get mocked or dismissed, the instinct is to run away, and to separate yourself from them and thus the world, to be defiant and embrace the alienation, to be a misanthrope. Basically, to be Morrissey. But I don't believe the river of ideas runs where loneliness lives. Isolation is desolation.

It is barrenness. Spending your creative days only there will leave your inner maker dying of thirst.

YOU'VE GOT TO GO DEEP

Paying attention to the things that thrilled me set me on a path that eventually led to cosplay. Engaging with my environment opened my eyes to the never-ending flow of ideas. But there's even another way to find inspiration, one that I have leaned on more and more as I've gotten older and more experienced: digging right through the bottom of the rabbit hole, by which I mean, going as deep as humanly possible on something you care greatly about, something you can't stop thinking about. This is how I cultivated my other great creative outlet and obsession: prop replication.

I've made no secret of my love for film and cinema as a medium. On my Mount Rushmore of cinematic influences, Stanley Kubrick sits squarely up there next to Ridley Scott, Terry Gilliam, and Guillermo del Toro. He is one of the directors by whom I am most inspired. The biting social commentary, the deep abiding sense of humor, the love of people with all of their foibles and flaws, that underlies each of his films has always resonated with me.

In 2013, I went to LACMA, the Los Angeles County Museum of Art, to see an exhibit about Stanley Kubrick. The exhibit was put together in close cooperation with his family, which, I hoped, meant it would be the kind of immersive, down-the-rabbit-hole experience I look for with a subject for which I have great passion.

I was wrong. It was so much better.

The Kubrick exhibit at LACMA was nothing short of a revelation. It contained a treasure trove of material spanning the director's entire career. There were pieces of every part of his filmmaking process: scripts with handwritten annotations by Stanley;

Deceptively simple-looking, Major Kong's survival pack was incredibly compelling to see in person.

card catalogs full of research notes on the life of Napoleon; costumes and props; cameras and lenses; production drawings, miniature sets, and incredible behind-the-scenes footage. For anyone who loves Kubrickiana, it was transformational.

On the day, I got to the museum early to give myself plenty of time to take it all in. I walked slowly and deliberately through the entire exhibit. I made sure to examine every photo, absorb the detail on every miniature, scrutinize every frame of available footage. All of it was thrilling. But one prop—one *singular* prop—left my inner maker completely stricken. I had just come around the corner from spending way too long ogling a display of Stanley's personal camera equipment (the lenses from *Barry Lyndon*!), to spy a large vitrine with a huge amount of paraphernalia from *Dr. Strangelove*, and right there in the middle was Major Kong's survival pack—a small, but fascinating prop from this wonderfully absurd film.

What's in the survival pack? Here's what Major Kong says when we see it for the first and only time toward the end of the film:

"Survival Kit contents check. In them you will find: one .45 caliber automatic; two boxes of ammunition; four day's concentrated emergency rations, one drug issue containing antibiotics, morphine, vitamin pills, pep pills, sleeping pills, tranquilizer pills; one miniature combination Roo-shan phrase book and Bible, one hundred dollars in rubles; one hundred dollars in gold; nine packs of chewing gum; one issue of prophylactics; three lipsticks; three pair of nylon stockings—shoot, a fella could have a pretty good weekend in Vegas with all that stuff . . ."

What a list!

The fact was that up until that moment I'd only ever seen Major Kong's survival pack as a narrative device. Now, sitting in front of me, seeing it as an actual object (with other objects inside it, no less) sent a huge thrill through me, one that I had been familiar with since my late teens and early twenties, when I was living in Brooklyn and spending time around the NYU film school, making sets and props for friends' student films. Whenever I got this feeling, it compelled me to ask myself: "What if I made one of those?" And the answer was always "*If?!?*" Once the idea comes to me, I have to make it, even if it takes years to start or to finish.

As I'll say many times in this book, I enjoy projects with many facets and high levels of complexity. This prop, this Kubrickian gold mine, was exactly my kind of project. Through contacts in Los Angeles, I was given permission to visit the exhibit again a month later, the day after it closed, where the curator allowed me to put on white cotton gloves and examine and make measurements of each part and piece of what remained of Major Kong's survival pack. And what remained turned out not to be that much. Maybe only about 20 percent of the listed contents survived. Clearly I had work to do.

What that work required, fundamentally, was to go deeper. The idea to replicate the survival pack had come to me as a product of going deep into the Kubrick exhibit. To actually make the pack would demand that I go deep on the pack itself.

But what does that mean, to go deep? As a maker, it means interrogating your interest in something and deconstructing the thrill it gives you. It means understanding why this thing that has captured your attention has not let go, and what about it keeps bringing you back. It means giving yourself over to your obsession.

I asked the director Guillermo del Toro once, if he thought there was a commonality to all great films, something that linked them to each other. He said you can never know from the inside if a movie is going to be great, but you can be sure that all great movies have at least one champion. Usually the director, but not in every case. A champion who lives, eats, sleeps, and breathes the film into existence, using all their passion, creativity, and obsession.

And yet, as a society we have a very suspicious view of obsession. In youth, as well as in adults, it is often considered a vice, a burden, an affliction. It gets hyphenated and transformed into a disorder by cynical armchair diagnosticians who have no problem casually conflating real, serious conditions like OCD with focused passion and conviction. People who are obsessive about things—about *any*thing, really—are crazy or addicted or out of their minds. We can't even countenance the idea that someone could be obsessed with something and be of sound mind. That is a shame, because when it comes to creativity, when it comes to making things, when it comes to success at anything, obsession is often the seed of real excellence. It inspires new ideas, it

1.

2.

3.

Rick Deckard's PKD Blaster: Version 1 (1987), Version 2 (1996), Version 3 (2008).

demands they be brought to fruition with exacting care, and it drives them to completion. With 80 percent of Kong's survival pack vanished to time, only obsession was going to push me to the places I needed to go to get it done at all, let alone get it done right.

My friend Bill Doran knows all about that. Like me, he is a prop maker and a huge cosplayer. He and his wife, Brittany, turned their very personal obsession with props and cosplay into an entire prop-making business, called Punished Props, complete with hugely popular video tutorials on their YouTube channel. For Bill, obsession is as much about inspiration as it is a motivating force against physics-related structural failures and the momentum-killing properties of indecision.

"No matter what you're making, no matter how good you are, you're gonna run into a thing that you don't know how to do or something goes wrong with your materials or you're running out of time, or whatever. And if you're not devoted to that thing, if you're not completely obsessed with that thing, you will stop," Bill said as we talked about the first big thing he ever made based purely on obsession—the armor for Commander Shepard from the awesome third-person shooter video game, *Mass Effect*. "But if I'm way into it and I'm super stoked, that thing right there, I won't be complete until it's built, nothing can stop me from making it. At all."

That's exactly how I felt about Kong's survival pack. And how I felt about every prop I've ever made for myself, beginning with the very first one: Rick Deckard's blaster from *Blade Runner.* I spent the better part of thirty years perfecting the fabrication of this prop. In fact, I've made three different versions since I first saw *Blade Runner* in 1985, each one better made and closer to the real thing as I got more experienced in the skills required to make a prop weapon like Deckard's sidearm.

In those thirty years, I moved from New York to San Francisco, I worked for numerous theater companies as a set builder, I worked on hundreds of commercials as a prop maker for Jamie Hyneman's and others' shops, I worked on over a dozen movies as a model maker for Industrial Light & Magic, I got married, I had kids, I made a TV show with Jamie for fourteen years, I got married again, and at no point did any of those events even come close to pushing the *Blade Runner* blaster out of my mind. In the back of my head, I was never really *not* working on it. If I wasn't actually fabricating components, I was rewatching parts of *Blade Runner* where it featured prominently, or I was researching gun-manufacturing techniques online, or I was reaching out to people who might know people who might know someone who saw or worked with the original prop, who might be willing to tell me which two guns were cobbled together to make this amazing piece of sci-fi weaponry, so that the last version I made was as perfectly identical to the original as I could make it.

This was the brand of obsession that brought me to the survival pack among all the other paraphernalia at the Kubrick exhibit. And it was this brand of obsession that *I* brought to its replication.

As I delved into the details, a meta-question began to form in my head: Why is the scene with the survival pack in the film in the first place? What does it tell us that was important enough to Kubrick to require its inclusion? For the uninitiated, *Dr. Strangelove* is an absurdist narrative about a nuclear misunderstanding that brings about the end of the world. A rogue general convinced that the Russians want to steal his precious bodily fluids sets off a chain of events that leads American bombers into Russia to drop our nuclear arsenal. The powers that be scramble to fix this and are able to recall all of the bombers save for one that's been shot at and damaged, such that it can only fly a couple hundred feet above the ground—making it conveniently invisible to both American and Russian radar. As our intrepid bomber crew, led by Major "King" Kong, limps steadfastly toward its final destination, the crew executes all of the operations of what will likely be their final mission, one of which is making sure that their survival packs are in order should they have to ditch out of the plane. Major Kong, played by classic Hollywood character actor Slim Pickens, reads out the contents of this survival pack on the radio, over shots of his crew at their stations in the plane following dutifully along. It's a weird and lovely break in the tense narrative, a refreshing breath that happens toward the end of the film. It's humorous and also deeply sad at the same time. You're watching this group of men heading toward their almost certain destruction, counting packs of chewing gum and pantyhose, executing their duty with a calm professionalism that is singular in the film.

I have come to conclude that the fact the bomber crew are the most competent and professional characters in the film is no accident. I think Kubrick wants us to understand that the tragedy of war is that it's often envisioned by idiots and executed by professionals. He's also clearly a fan of banal conversations; they exist in some form in nearly every one of his films. They are exchanges that don't necessarily propel the plot but that give deep insight into the world of the film.

The second question I had was: Why these items? As part of my research, I collected and chronicled many different types of survival packs given to pilots and airmen from World War II up to and past the point at which *Dr. Strangelove* is supposed to be taking place, and Kubrick's production team did their job very well. Most all of the items listed by Major Kong would've absolutely been included in any survival pack of the time. But things get weird when we get to items like the lipsticks, the prophylactics, and the nylon stockings, we're waaaay out away from anything that would have been normally included in the real thing.*

So then the question becomes: What is Kubrick saying by including them? Personally, I think he's adding a new narrative to the bomber crew's potential survival. He's letting us know in the absurdist version of the world he's created that American airmen, in order to escape, may need to use gold to bribe their way through Russian men, or lipsticks and stockings to bribe their way past Russian *women.* With just the addition of those few items, Kubrick's brain is 3-D-printing another layer into his absurdist vision and another subplot (if they *do* survive, how will they escape?) deep into our heads.

These questions I asked of myself were not random. They arose from my deep examination of a single, seemingly banal prop at the most minute detail possible. I found answers in the depths of my exploration, answers that also helped inform how I would replicate the pack and its contents, what materials I would use, and why they were necessary. The whole idea of replicating the survival pack emerged from my obsession with fantastical, filmic narratives and a deep interest in understanding them and what they mean to me. By traveling down the path of research and replication, I gained a new, deeper insight into a filmmaker whose work I find endlessly fascinating, endlessly inspiring. So much so, that I would seize upon many more Kubrick replica ideas in the years to come.

EMBRACE YOUR BRAIN

I have always found that to make anything great requires a good idea that is approached with a genuine regard for excellence and honesty. For me, those ideas are most often born from the rigor-

---

* Soldiers were issued condoms, but they weren't, as far as I could determine, ever included in a survival pack.

ous examination of myself, my world, my surroundings, my culture, and my interests. When an encounter with another's work moves me at a deep level—whether it's a character in a story or an object in a film—my desire to embody that character or replicate that object is really just an attempt to understand and unpack why it has moved me and then to capture the story of that moment of inspiration in physical form. But I also recognize that that process of inspiration and ideation is unique for everyone and every creative pursuit.

How I have come up with my stories, how I get ideas for what to make, often comes from my love of movies. But your ideas can come from anywhere. They are out there, floating everywhere. It will be your interest and obsession that create the gravity that draws them to you and then makes them yours. If you can feel that draw, that attraction, and then something catches your interest as a result, PAY ATTENTION TO IT. Being attuned to those pings of interest is the duty of a creator, whether you're a scientist coming up with a hypothesis, or an artist with a blank canvas in front of you, or a troubadour with a quiet guitar in your hands. We all have brains, and the ability to do remarkable things with them, but what we do with them is up to each of us.

Beyond that, there is no magic formula for getting started, I promise. It merely requires that you participate in your world, that you pay attention to what interests you, that you follow the thrills they produce, and that you never be afraid to go deep on them, to obsess over them, to dig through the bottom of the rabbit hole, if necessary, to find that great idea that has been waiting there for you all along.

LISTS

When I was younger, if you'd told me to make a to-do list before I embarked upon a project, I'd have rejected the idea out of hand. List making was the death of creativity! It was a stultifying tool of the ordered, methodical, measured, boring universe. Creativity was the rainbow connection. It was an electrifying, soul-flight on the wings of imagination . . . or something like that. To clip those wings with something so basic and banal as a list was not just a sin, it was counterproductive. List making slowed me down. It was nothing more than an obstacle between the moment of inspiration and the process of creation.

I think a lot of creative people look at planning tools like list making that way at some point in their lives. Planning is for parents. Lists are for accountants and teachers and bureaucrats and all the other agents of creative repression! But here's the thing: lists aren't external to the creative process, they are intrinsic to it. They are a natural part of any project of scale, whether we like it or not. Even when I was a young found-object artist living in Brooklyn after high school, and I refused to make physical lists at the start of any of my projects on principle, my brain still tried to catalog all the stuff in front of me every time I set to work, and what is a catalog but a systematized list?

Now I love lists. I like long detailed lists. I like big unruly lists. I like sorting unsorted lists into outline form, then separating out their topics into lists of their own. Every single project I do involves the making of lists. I make them for organization, of course, but I also make them for assessment, for momentum, as a stress reliever, and, counterintuitively (at least to my eighteen-year-old self), as a means to improve my creativity and free my thinking. There are daily lists, there are project lists. There are "things to order" lists. I make lists of pieces of research that I want together, lists of people I am collaborating with, and what they need from me to support their part of the project. I make lists of things I need to purchase, things I need to find, and when all of those objects are going to get to me. And hopefully, finally, there are "homestretch" lists, that tell me I'm reaching the end. This sounds like Dr. Seuss, I know, and that's not such a silly comparison, because if lists do anything, they give rhyme and reason to any project, big or small.

I first recognized my natural affinity for lists all the way back in 1979, when I was twelve years old and my parents got cable television. Until then, like everyone my age, we only had six or seven channels to watch. There were the three big networks—CBS, NBC, ABC—and PBS, and whatever drek was in the UHF wasteland that Weird Al Yankovic would make popular ten years later with his movie of the same name.* But CABLE! Oh my goodness. Right off the bat, there were fifty to sixty new channels to explore. And like any preadolescent suburban kid, I mined their programming for anything and everything I could not otherwise find on network television.

* See, I just made a list without even trying!

It didn't take long to strike gold. I found it amongst the "chicken shit a-holes" (still one of my favorite movie insults ever) of *The China Syndrome*, the 1979 nuclear disaster film starring Jane Fonda and Michael Douglas that came out on cable later that year. It wasn't the tense depiction of nuclear holocaust on TV that captured my imagination, however. It was the CURSING! For a twelve-year-old, seeing people curse on TV was way more of a revelation than the prospect of atomic annihilation. Whereas network television produced a sanitized version of reality that never really resonated with me even then, cable TV was like a window on to the world as it actually existed. This is how the older kids on the playgrounds talked when adults were out of earshot. This is how my parents' friends talked at our monthly dinner parties when they thought all the kids were asleep. It's how my parents talked in front of me (bohemians!). The people on cable TV, the characters who cursed, were just more real.

And then there was George Carlin.

The year before, Carlin produced the second of what would eventually be fourteen incredible stand-up comedy specials for a fledgling cable network called HBO. "George Carlin: Again!" was also the second of three straight specials that would include, famously, extended versions of a track from his 1972 *Class Clown* album called "Seven Words You Can Never Say on Television." I am not going to name those words here—that's why the internet was invented—but you should know that hearing the seven dirtiest curse words that supposedly should never be uttered on television, being repeated over and over again to uproarious laughter and applause from the audience, was intoxicating to the twelve-year-old version of me.

Then it got even better. He confided to his audience that those seven dirty words were just "my original list, I knew it wasn't complete, but it was a starter set." Then he rattled off twenty-five new ones to round it out. I'd never heard of most of these terrible/wonderful things. My mind reeled. Such fabulous filth! I wanted this list. But that was easier said than done.

This was well before the days of DVR, and the mass-market version of the humble VCR machine was only a few years old at the time, and prohibitively expensive. If I wanted that list of curse words, I would just have to wait until Carlin's special aired again and be ready with pen and paper in hand to transcribe his recitation of raunch as fast as I could. It wasn't easy. He spit the last twenty curse words out in about thirty seconds, many of them obscured by audience laughter. I had to watch the special at least a half-dozen times over the period of a month or so to make sure I got them all—cross-checking with previous lists as I went to make sure I'd jotted them down correctly. Sure, the list was fun because of its naughtiness, but I was most drawn to the task because it was the most complete version of anything I could have conceived of as a kid. There was no way I could acquire all the LEGOs or all the *Star Wars* action figures, which were my two other favorite pastimes just a year earlier, in 1978 BC (before cable). Indeed, this was my first true *collection*, and I can see now, looking back, that this was the moment the completist in me came out.

YOU COMPLETE ME

My completist tendencies had bubbled to the surface once before, I just didn't know that's what was happening. When I was younger, my older sister Kris had a huge 5,000 ml Erlenmeyer flask full of pennies that went all the way back to the 1930s. When she was at school or out with her friends, I would sneak into her room, dump all the pennies out onto the carpet and sort them into piles by year, looking for the rare (to me) wheat pennies and hoping for an unbroken chronology of coins. Her collection was so cool, except there were always gaps between years, which was not cool. Each gap I spotted made me more vexed, because it meant her collection was that much further from completion, and that simply would not do.

My preoccupation with completeness manifested itself constantly throughout my teenage years and has continued well into adulthood, again most often with film props. Specifically, and conveniently, with film props I was *not* being paid to create. When I become fascinated by a particular prop, my interest never stops with just that object. I want to know everything that could orbit around that prop—the world in which it exists, the things with which it interacted.

For longer than I can remember, I have been enamored with the space suits in Stanley Kubrick's *2001: A Space Odyssey.* In 2015, I finally completed a replica of one of them. Not one of the vivid, primary-colored suits worn by the astronauts on the *Discovery,* however. Those have been replicated by enthusiasts many times over. I chose a suit that I'd never seen replicated before: Heywood Floyd's silver Clavius Base suits from early in the film.

The silver Clavius suits, which had their own integrated cooling systems, featured polished white helmets and highly detailed front and backpacks. Developed in conjunction with engineers and scientists from the American space program, they looked as legit as any space suit ever in film. To begin, I commissioned the bulk of the suit from Mike Scott, who makes brilliant replicas of the *2001* suits and helmets, though up to that point he'd never tackled the silver suit. Then, as his parts of the suit came together, I built its backpack, an internal ice-water-powered cooling system, and most of the machined aluminum parts that gave the suit its realistic and futuristic feel.

Chris Hadfield, former commander of the International Space Station, and I walking the floor at the 2015 San Diego Comic-Con in our Clavius suits. Oh, and the smiling man in the rumpled shirt watching us pass—that's The Martian *author, Andy Weir.*

The whole process took nearly four years, hundreds of hours of research, and dozens of lists. I pored over the tiniest of details. I examined Kubrick's own research and how it dovetailed with NASA's ideas about what space travel and its accessories might look like. I wanted to get a sense for the logic that led to this design, so if there were pieces I couldn't get enough detail for by watching the film a thousand times, I at least had a solid theoretical foundation from which to replicate them. As I did that though—watch the film a thousand times—I became tangentially obsessed with a different object in Heywood Floyd's universe: his lunch box.

As Floyd and his colleagues make their way on the moonbus to see the Monolith at Clavius Base, they eat lunch from this cool-looking, oblong, octagonal white box with big thick clips on the side and an equally geometric, removable lid. As I stared at it over and over again with each viewing, the container seemed to me to quite clearly be a found object—that is, an existing product that prop makers for *2001* might have added specialized clamps to and repurposed to be a lunch box. I still believe this. Yet no amount of online searching or archival research yielded anything remotely close. I looked for early-1960s sharps containers, bicycle and motorcycle panniers, lunch boxes. I made a list of all the things I thought it could be, then did two hundred or three hundred separate eBay searches, using different combinations of keywords, to see if I could get close, and still nothing.

The completist in me would not let this stand. Once I caught the bug to fill out the frame of the *2001* Heywood Floyd tableau I had in my head, I could not shake off my desire, *my need*, to obtain one of these lunch boxes. And if I could not buy it, well then I was just going to have to build my own from scratch, which is what I did.

But I didn't stop there. When Floyd opens the lunch box, there's a piece of paper that is briefly visible. I figured that was likely a requisition form for the contents of the lunch box. Before Floyd set out on the shuttle to Clavius Base, a list of sandwiches would've been put forth as suggestions for the trip, and fulfilled by NASA. I decided I had to replicate this requisition form, as well. I perused hundreds of NASA documents, with the deep satisfaction only a completist could fully appreciate, until I was confident that

Heywood Floyd's lunch box in my workshop.

I knew what this requisition form might look like and how I might reproduce it. For the finishing touch, I ran it through a fairly harsh series of Photoshop filters to make it look as if it had been copied many, many times. It was the additional bit of authenticity I felt the form needed to feel like it was a regular piece of the world Heywood Floyd inhabited.

Completism and list making create a feedback loop: completism demands list making in order to be successful; list making begets completism, because why would you make a list if it didn't contain everything you needed? This feedback loop can flip negative if it becomes the sum and substance of all your creative effort. Just think of all the perfectionists you know who can't get

PAGE ____ OF ____ PAGES

NATIONAL AERONAUTICS AND SPACE ADMINISTRATION

CL/B Rocket Bus Provision Requisition Short Form VII-24

Form Approved O.M.B. No. 2700-0003

Date(s) of Mission: Wednesday April 26, 2001- Friday April 27, 2001

Mission Name: CL/E Quarrentine Tycho Reconnaissance

Mission Parameters: 3 man team to Tycho

No. of Personnel (excl. Pilot & co-Pilot): 4.

Authorized Supervisor (signature required)

List Personnel:
Dr. Floyd
Dr. Halvorsen
Dr. William "Bill" Michaels

Destination and Duration: Land at Tycho Creter at 5:45 EDT after transit time of 25 hours (approximate). Dr. Floyd will suprvise team upon arrival

Cost: $ NA · $ NA · $ NA · $ NA · $ XXXXXXX

PROVISIONS REQUESTED

Type	No.	Type	No.
Ham Sandwich	14		
Turkey Sandwich	11		
Tuna Sandwich	10		
Mayonnaise	2		
Trail Mix	3		
Please pack coffee if available. Dr. Floyd authorizes use of non-pressurized thermos for trip			

Abbreviations:
(A/S) Artificial Sweetener
(B) Beverage
(FF) Fresh Food
(IM) Intermediate Moisture
(I) Irradiated
(NF) Natural Form
(R) Rehydratable
(T) Thermostabilized

ROCKET BUS SHORT FORM PROVISIONS

Sandwiches: Ham (NF), Tuna (NF), Turkey (NF), Roast Beef (NF), Cucumber (NF), Cheese (NF)
Condiments: Mayonnaise (NF), Mustard (NF), Relish (FF), Barbecue Sauce (IM)
Dressings: Italian (IM), Thousand Island (IM), Tahini (IM), Blue Cheese (IM)
Cheeses: Cheddar (T), Swiss (T), American (T)

Fruits: Apple, Granny Smith (FF), Apricots, Dried (IM), Applesauce (T), Grapes (FF), Orange (FF), Pears, Diced (T), Peaches, Dried (IM), Peaches, Diced (IM), Pineapple (T), Strawberries (FF), Trail Mix (IM)
Meats: Beef (R), Chicken (R), Pork (R)
Jelly: Apple (T), Grape (T)

Juices: Orange (B), Tomato (B), Pineapple (B), Apple (B), Carrot (B)
Cereal: Bran Chex (R), Cornflakes (R), Granola (R), Granola w/berries (R), Grits (R), Oatmeal w/brown sugar (R), Oatmeal w/raisins (R), Rice Crispies (R)
Grains: Rice (R), Quinoa (R), Polenta (R)

APPROVED — PROCUREMENT OFFICER — DATE

(Please note: CL/B makes every effort to keep proper foodstuff and provisions available. However, unanticipated interruptions to resupply missions can affect current stack of provisions on-hand. CL/B may make substitutions.)

Baseline Plan Identification (Col. 7b & 7d); Revision No. ________, Dated ________

NASA FORM 737H AUG 00 PREVIOUS EDITIONS ARE OBSOLCTE

This is like my dream come true—a prop replica that is also a list!

started unless they have everything exactly right, in exactly the right order, down to the most granular level.

My colleague and friend Jen Schachter battled this tendency early in her career. Jen is the artist-in-residence at a maker space in Baltimore called Open Works, where she focuses on the intersection of the maker movement with education, employment, and equality. "I think for a long time it was something that I struggled with, not just with creative work, but with my academic work, too," she told me. "Any time I needed to write a paper or do a presentation, I was so attached to it being, not to say perfect because I never thought I was perfect, but that it had to be done a certain way. I struggled with it because everything took me longer than it should take me to do things. It can really hold you back."

Completism can certainly tip toward perfectionism and into a negative feedback loop, but there is also great utility in it—a *positive* feedback loop—when it comes to list making as a planning tool. It ensures that you capture the totality of your project: the materials you will need; the quantities required; the collaborators you might need to coordinate with; the steps it will take to put those materials together. It's all there in front of you on the page, in one place. Far from inhibiting creativity or obstructing the making process, lists like these actually unlock your creative potential because they free up all that brainpower you would have otherwise been using to remember all this information.

TAMING THE BEAST

I learned this with my very first professional list back in 1991, when I was twenty-four years old and working with a group of friends in the San Francisco theater scene trying to get our own theater company off the ground. We were five in number, and we called ourselves V MAJEC, a play on our names: V being the

Roman numeral for five, MAJEC being comprised of all of our first initials. It was one of those ideas that sounds really cool when you're twenty-four years old, but by the time you hit thirty, you wish you'd let it slip through your fingers when it came floating by. Still, we were talented, earnest, idealistic, and super ambitious, and we dove into the enterprise head- and heart first.

I was in charge of construction. My job was to put together all our equipment, build the sets, the props, the lighting truss, and pretty much handle anything that required a maker's instincts. Even for a tiny company like ours, this was no small task. On the eve of our first big gig, I spent six hours with my friend Steve at his house on Rivoli Street in San Francisco's Cole Valley, figuring out the full extent of the crushing amount of work to do before opening night, to make sure we left nothing undone. As we talked and went scene by scene, department by department, I began furiously making lists, and more lists, until I was making lists of lists that took up several pages. It was the middle of the night, I'd compiled what we needed, yet I still didn't feel like I had my head wrapped around all the work. That would not do, so I decided to consolidate everything.

I spread all the lists out in front of me on the living room floor and meticulously copied their contents onto a single page in the tiniest writing I could manage, so that I could "see" all the work we had to do, all at once. This was an important moment. The moment at which our first show took real shape in my head. This wasn't the original goal of transcribing my lists, mind you. There was no conscious intent on my part to make a master list on a single page in order to tame this beast. And yet, with the entirety of the show's elements in front of me, that's exactly what happened. Once I could see the whole thing, the show became a singular, manageable project in my head instead of an endless series of sep-

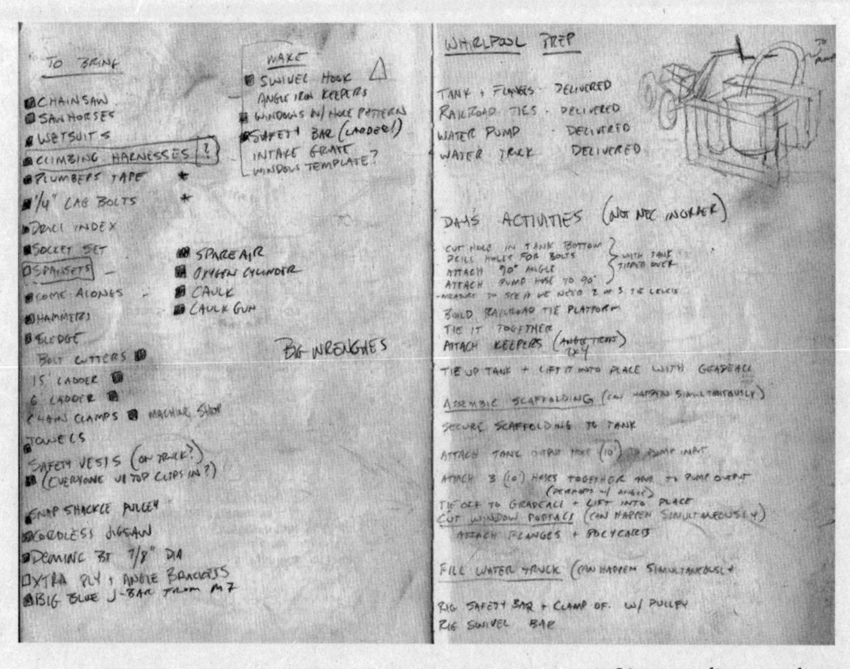

Lists upon lists upon lists.

arate, overwhelming tasks. This, I quickly understood, is the abiding power of a well-made list. When your project feels like a lion that needs to be tamed, a comprehensive checklist can be both the whip and the chair.

This is as true for individual creative projects as it is for entire industries. List making has transformed many of them, right along with the lives of the people who interact with them.* In medicine, simple checklists, like the Apgar score, have saved countless babies' lives, and saved countless hospitals hundreds of millions of dollars in waste. In aviation, compiling reams of instructions has become a list-making science of precise font-size, field-tested

* Atul Gawande wrote a terrific book about this called *The Checklist Manifesto*, and it's worth reading.

readability, and total items allowed per page. Not only has this made takeoffs and landings more efficient, but it's also made it easier for people in malfunctioning planes to follow multivariant emergency procedures through shifting, uncertain circumstances and get everyone safely on the ground as quickly as possible. In each case, the thing lists solve for—the beast they tame—is complexity.

The maker in me knows that this is where lists really shine, that it is their capacity for simplifying the complex that sets them apart from all other planning tools. Not just at the beginning of a project, either, but at every step along the creative process, because no matter how exacting the list you make at the outset, there will always be things that you missed, or, more frequently, that change. It's like trying to measure a coastline: it's fractal. The closer you look, the longer it gets, as will the lists you create to incorporate all the new, changed information. To that extent, when it comes to measuring my progress, the lists aren't so much guides as they are maps through an ever-changing landscape. And anyone who knows my work knows how much I like elaborate and knotty projects. Don't get me wrong, I've made plenty of simple props over the years, like a perfect Indiana Jones Fertility Idol, or the case that houses Hellboy's gun, The Good Samaritan. But complexity is what draws me in like a tractor beam. It drew me right to another *Hellboy* prop, in fact: Rasputin's Mecha Glove.

In the beginning of *Hellboy*, Rasputin, the film's villain, gives birth to Hellboy and delivers him from a shadow realm to this mortal plane using an incredibly intricate device known as the Mecha Glove. I obsessed about this prop for a long time. When I finally met the film's creator, Guillermo del Toro, in 2009, I could have talked to him about any one of a half dozen of his amazing films—*Pan's Labyrinth*, *Mimic*, *Cronos*, *Blade*—but instead I

Mecha Glove diagram.

gushed about how much I loved the Mecha Glove. It was a brilliant artistic feat, a playground of complexity, I told him.* And because Guillermo is basically a giant beating heart wrapped in a beard, he immediately put me in touch with Mike Elizalde, who built the original hand . . . and still had it.

As soon as I could, I visited Mike and the Mecha Glove in his shop Spectral Motion in Los Angeles. Mike graciously supplied

* . . . I think. I may have blacked out during our conversation!

INTERIOR DETAIL FOR TUBES

Finger tubes are 1.5" diameter and wound with .035" magnet winding wire. The base is knurled, the shape is slightly conical, and there are 6 small and 1 large brass screw in the tip.

Tube 1 is 1" diameter and has the guts to a real radio tube inside.

Tube 2 is 1.375" diameter and, like Tube 1, has the guts to a real radio tube inside. Both tubes look like the guts are identical to each other.

Tube 3 is 1" diameter, and has a triangularly folded etched brass construction inside. The brass is stock from model trains.

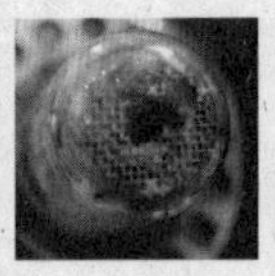

Tube 4 is 1.25" diameter, and also has model-train brass inside, but a different kind to Tube 3. It's in 3 tiers and also has some screen covering the end.

Tube 5 and 6 are both 1.25" diameter and is identical in dimension to Tube 4. That means that there are 3 tubes of this dimension on each hand. It is filled with a custom sculpted and cast mummified frog, sitting in liquid.

The cages are 1.3125" diameter and there are 6 of them per hand, 5 shorties and 1 long one.

Tube 7 used to be filled with a ball with some RHOD type designs on it, but for filming had a metallic-like liquid and a pump built in that moved it around. I'm not exactly sure how. It's 2" in diameter.

me with the proper references, measurements, and advice to make my own Mecha Glove if I wanted. Well, don't threaten me with a good time!

Composed of more than six hundred separate pieces and parts, my replica Mecha Glove took almost four years to finish and required easily hundreds of lists. There were lists of things to build, lists of hardware to obtain, lists of things to research, lists of people to press for information.* I made lists of problems I'd yet to solve, lists of words to guide my choices when I had to guess about something. I made lists of companies that could manufacture specific parts for which I had neither the equipment nor the

OPPOSITE: *My apologies to the late Senator Ted Stevens, but this chart from my Mecha Glove build is the real series of tubes.*

* I make this list with every complex project, and lean on it often. On a recent space suit build, I texted astronaut Mike Massimino every day to ask what he kept in various pockets of his orange shuttle suit.

expertise.* I even drew some of my lists and made pictorial versions of them. With so many different pieces, no two the same, this became a critical way to tame the beast so that I could understand it.

Regardless, whatever form my lists took, their job was the same: to constrain and contain the unwieldy, to organize the chaotic, to simplify the complex. In fact, the level of complexity I enjoy in my projects has increased over the course of my life, because my capacity for parsing it has also increased, thanks almost entirely to these very lists. And as time-consuming as these lists can be to create for the listically disinclined, from a planning perspective, I have yet to find anything else that can deliver this kind of effectiveness for your project and growth for your own capacity as a maker.

---

* I commissioned all the glass for the finger tubes from an amazing company called Adams & Chittenden, which makes all the custom glass for the labs at UC Berkeley.

CHECKBOXES

Making useful lists has been a lifelong process of refinement. This began early, with the brute force of youthful enthusiasm, gathering as much information on a given subject as I could. When I got a little older, I added an invaluable layer of finesse to my list making thanks to a few years spent working in graphic design. The job of graphic design is to communicate important information as quickly and as efficiently as possible. If I wanted my lists to deliver the information I needed, when I needed it, and to not overwhelm me in the process, then they needed to be clear and concise. Graphic design helped me home in on lists in outline form and as diagrams with pictorial references to clean up the mess of some of my earlier list-making tendencies. Don't get me wrong: those first messy lists were still useful, as any list is better than no list. But the cleaner they got, the better they got, and the more productive I became. Then, in 1998, I joined Industrial Light & Magic (ILM) to work as a model maker on *Star Wars: Episode I—The Phantom Menace*, and my list making took a quantum leap. I discovered the checkbox.

When I arrived at ILM, the checkbox was already a piece of institutional practice. I noticed it one day early in my tenure, looking over the shoulder of my boss, Brian Gernand, as he went down his to-do list for me. Next to every item on his list, down the left-hand side, he'd drawn little boxes. Some of the boxes were empty, some were colored in, others were only partially filled in. I asked him about it when we were done and here is how he explained his process:

- ■ If a task was completed, he colored in the corresponding box on the list.
- ◪ If a task was halfway or mostly complete, he colored in half its box diagonally.
- ❑ If a task hadn't been started or measurable progress had yet to be achieved, that box stayed empty.

Brian is one of the best supervisors I have ever worked for. I've seen him manage anywhere from a half dozen to hundreds of builders in the ILM model shop. On a big project like a feature film, each one of those builders works from daily, weekly, and sometimes monthly to-do lists, for production periods lasting sometimes up to a couple of years. The number of tiny details captured in those lists is immense. On a *Star Wars* picture, it is positively gargantuan. It is easy to see how a supervisor, whose job it is to oversee all that, could drown in the details. And yet, this three-part checkbox technique allowed him to see instantly where he was in any project, at any given moment, on any given day.

The elegance and effectiveness of this planning system floored me, particularly when it came to evaluating the status of a project the further along it went. The value of a list is that it frees you up to think more creatively, by defining a project's scope and scale for you

You can see everything about your project in a list with checkboxes.

on the page, so your brain doesn't have to hold on to so much information. The beauty of the checkbox is that it does the same thing with regard to progress, allowing you to monitor the status of your project, without having to mentally keep track of everything.

I incorporated the checkbox into my process immediately and it changed my work practice at ILM overnight. Every day from that day forward, I would make a new daily list of goals complete with checkboxes, while at the same time keeping an eye on how that day's goals fit into the bigger picture. I'd make that day's list of goals by looking at yesterday's list, and transferring only the unstarted or partially completed items to today's list. It was a great way for me to wrap my head around the day, the week, the month, and my job in general. I became known for being so meticulous about tracking my progress through my lists, in fact, that it opened me up to the occasional prank from my coworkers. I'd come into the shop in the morning, ready to set myself to the task of creating that day's to-do list, only to find "uncompleted" items added to the previous day's list: "Buy Mike Lynch lunch"; "Give Brian G $10." Har har.

The power and importance of the checkbox for me simply cannot be overstated. On the one hand, it speaks, as I've said, to the completist in me. The best part of making a list is, you guessed it, crossing things off. But when you physically cross them out, like with a pen, you can make them harder to read, which destroys their informational value beyond that single project and, to me at least, makes the whole thing feel incomplete. The checkbox allowed me to cross something off my list, to see clearly *that* I'd crossed it off, and at the same time retain all its information while not also adding to the cognitive load of interpreting the list.

The checkbox also resolves some of the tension inherent to my physics-related approach to creativity. In my mind, a list is how I describe and understand the mass of a project, its overall size and the weight that it displaces in the world, but the checkbox can also describe the project's momentum. And momentum is key to finishing anything.

Momentum isn't just physical, though. It's mental, and for me it's also emotional. I gain so much energy from staring at a bunch of colored-in checkboxes on the left side of a list, that I've been known to add things I've already done to a list, just to have more checkboxes that are dark than are empty. That sense of forward progress keeps me enthusiastically plugging away at rudimentary, monotonous tasks as well as huge projects that seem like they might never end—and there were plenty of both during my time at Industrial Light & Magic. From the six-foot-tall, skyscraper-sized crane for a bank commercial that encompassed eighty hours of laser cutting and weeks of assembly to communicate its impossible scale, to building a Thermian docking station for

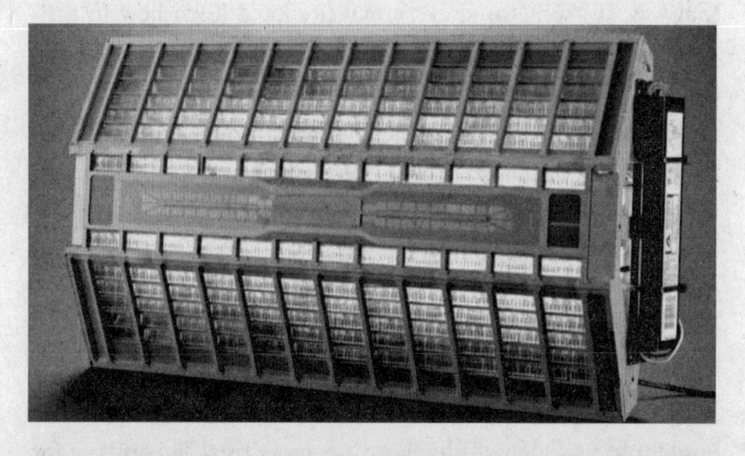

The Thermian docking station from Galaxy Quest.

Galaxy Quest composed of hundreds of backlit windows, each one with a tiny slide of Thermians behind it.

There's something about not just capturing and riding momentum in a project, but building more of it, that keeps me racing back to my shop in the morning, day after day, with my feet planted firmly on the ground, and my mind and my project pointed in the right direction. It sounds funny to say, but I've trained myself to be my own momentum propaganda machine that way. It's something every maker should learn how to activate for themselves, I think, because you can't count on external sources of motivation to be there when you've hit a wall with a project, or you're in the dead days halfway through. You will need to create your own motivation to keep going, and the momentum that springs from a checklist that is more filled in than not can be just the thing to fuel your fire.

HOW I MAKE LISTS

There is a famous Haitian proverb about overcoming obstacles: Beyond mountains, more mountains. One could easily apply that wisdom to making: within every list, more lists. This is true for every project, from the simplest to the most complex. This raises an important question for many makers as their work grows more ambitious. If a list is meant to tame your project, how do you tame a list that can be broken down to infinitely deeper and deeper levels? After a lifetime of trying to break this bronco and corral it into the barn, this is what I've come up with.

Step 1: The Brain Dump

My method of list making is not to make a single list per project, but rather to make a series of lists that helps me define it as it goes. The first series is a process of refinement whereby I begin to wrap my head around the scope of the project in front of me. This process of being able to understand the whole thing begins with a big brain dump. I will sit at my desk at home or my bench at the shop and just write out everything off the top of my head, willy-nilly. Anything I can think of, even tangentially related to the project, goes onto the page. It's a jumble, and that's the point.

The first pass at *anything* is always going to be a confusing mess. The story genius Andrew Stanton, who cowrote the *Toy Story* and *Monsters, Inc.* franchises and directed both *Finding Nemo* and *Finding Dory* for Pixar, talked to me about this first pass at listing out the component parts of a project. He was consulting with a group recently, and they were working on the early stages of a project and he said to them: "Can we just all agree right now, this is going to suck? Whatever we're talking about now, no matter how much we're getting excited, let's just all understand it's going to be a mess." They were shocked, wondering if he was insulting the project itself. He explained that no, he was simply letting them know that in a project with any amount of complexity, the early stages won't look at all like the later stages, and he wanted to take the pressure off any members of the group who may have thought that quality was the goal in the early stages.

The same applies to any making project. Mark Frauenfelder, the founding editor in chief of *Make:* magazine, insists that "you've got to do at least six iterations, minimum, of any project before it starts getting good enough to share it with other people." That very first iteration, what I call the brain dump, Mark calls "the quick and dirty stuff."

Here's what the quick and dirty stuff in my brain dump might look like if we were making, say, a ray gun:

- [ ] GUN
- [ ] STOCK
- [ ] SCOPE
- [ ] HOLSTER
- [ ] ELECTRONICS
- [ ] LIGHTS
- [ ] SOUND BOARD
- [ ] SOUND FILES
- [ ] BELT ATTACHMENT
- [ ] CASE
- [ ] GAUGE
- [ ] TRIGGER GUARD
- [ ] AMMO
- [ ] TRIGGER
- [ ] ADJUSTMENT DIALS

This is mostly a sacrificial list, because without an overarching order, or recognition of the many components within each item on the list, it's always going to be overwhelming and incomplete. Just like what's happening in my brain. Looking at a brain dump list like this gives me the same feeling I get when I have been away for a few days, and my kids have been staying at the house, and I walk into the kitchen for the first time. It's a disaster area. It's not unrecoverable, I know what needs to get done, but looking at the totality of tasks in front of me . . . I get overwhelmed by the sheer volume of it all. The brain dump is still a useful list, don't misunderstand, just not in its current form. It is simply the first step of the process.

Step 2: The Big Chunks

Step 2 is to take that massive list and start to carve it into manageable chunks. Looking at the brain dump list for our ray gun, I can already see that there are some broad categories to break out and then break down: the gun, the case, the holster, to name a few. So now, immediately after writing out the entire brain dump list, I start to rewrite it, this time in more of an outline form.

- ☐ GUN
  - ☐ GRIP FRAME
  - ☐ TRIGGER
  - ☐ TRIGGER GUARD
  - ☐ GAUGES
  - ☐ SCOPE
  - ☐ BODY
- ☐ GUN ELECTRONICS
  - ☐ LIGHTS
  - ☐ SOUND BOARD
  - ☐ SOUND FILES
  - ☐ SPEAKER
  - ☐ ACTIVATION CIRCUIT(S)
  - ☐ MICRO SWITCHES
  - ☐ VOLUME ADJUSTMENT
- ☐ GUN CASE
  - ☐ BOX
  - ☐ LABELS
  - ☐ LINING
  - ☐ INTERIOR CUTOUTS
  - ☐ INSTRUCTIONS
- ☐ HOLSTER
  - ☐ LEATHER BODY
  - ☐ METALLIC LABEL
  - ☐ BELT ATTACHMENT
  - ☐ BELT
  - ☐ BUCKLE
  - ☐ CLOSURE SNAP
  - ☐ GRAPHICS
- ☐ AMMO
  - ☐ BULLETS
  - ☐ AMMO BOX
  - ☐ GRAPHICS
  - ☐ POWER PACK BODY
  - ☐ POWER-UP ELECTRONICS

Step 3: The Medium Chunks

With the big chunks laid out and preliminarily itemized, I reorganize the components into medium chunks by subcategory. Let's take the top-level "Gun" chunk, for instance. I can immediately see, now that it's been broken down, that there are three primary subcategories into which these discrete tasks can be grouped. There is 1) the gun body itself; 2) the electronics that go into it; and 3) the graphics that go around it.

- [ ] GUN
  - [ ] MAIN BODY
    - [ ] 3-D PRINTING
    - [ ] BRASS DETAILS
    - [ ] MACHINED ALUMINUM
  - [ ] GRIP FRAME (ALUMINUM)
    - [ ] ALUMINUM
    - [ ] MACHINED VOID FOR SWITCHES
  - [ ] GRIP SIDES
    - [ ] SCREW BOSSES FOR ATTACHMENTS
    - [ ] PURCHASE PAN-HEADED STAINLESS SCREWS
    - [ ] THREADED INSERTS FOR SCREWS
    - [ ] CHECKERING/GRAPHICS (BRASS?)
  - [ ] TRIGGER
    - [ ] TRIGGER MECHANICS
      - [ ] SPRINGS
      - [ ] WIRING
      - [ ] MICROSWITCH
      - [ ] HAMMER MOVEMENT
    - [ ] TRIGGER GUARD (STEEL?)
      - [ ] FRONT & BACK MOUNT
  - [ ] LEFT SIDE GAUGE
    - [ ] REMOVABLE FOR DETAILING
    - [ ] BEZEL
    - [ ] GRAPHICS
    - [ ] DIAL NEEDLE
      - [ ] SERVO FOR ACTIVATION (MICRO)
    - [ ] ADJUSTMENT

- ☐ SCOPE
  - ☐ LENSING
  - ☐ BODY
  - ☐ MOUNTS
  - ☐ ELECTRONICS
    - ☐ LIGHTS
    - ☐ SWITCH
    - ☐ WIRING INTO BODY
- ☐ ADJUSTMENT DIALS
  - ☐ KNURLED ALUMINUM (LATHE)
  - ☐ GRAPHICS
  - ☐ POTENTIOMETERS
  - ☐ ATTACHMENT
- ☐ PAINT
  - ☐ SCOPE (BLACK)
  - ☐ GRIPS (VARNISH)
  - ☐ MAIN BODY
  - ☐ WEATHERING
    - ☐ CHIPPED PAINT
    - ☐ RUST
    - ☐ DIRT
    - ☐ VERDIGRIS
- ☐ ELECTRONICS
  - ☐ ARDUINO BOARD
  - ☐ BATTERY COMPARTMENT
  - ☐ INDICATOR LEDS
  - ☐ SOUND BOARD
    - ☐ FILES
    - ☐ ACTIVATION SWITCH(ES)
    - ☐ SPEAKER
- ☐ GRAPHICS
  - ☐ GAUGE GRAPHICS
  - ☐ BRASS DETAILS (ETCHED)
  - ☐ DECALS (ORDER)

This is where the real scale of the project begins to reveal itself. This is also where I begin to feel a sense of relaxation come over me, as the formerly incoherent brain dump I started with coalesces into an ordered, coherent order of operations. I do this with way more than just building projects. I did it with this book, I do it with things like moving house, throwing parties, and sending gifts. I find making lists to be one of the greatest stress reducers besides meditation. Be careful though, the itemizing of list detail risks becoming a self-perpetuating process: the more items you think of, the more new ones will come to mind. You can lose your whole life inside these lists, making them ever more detailed, so at a certain point it's time to just begin.

Step 4: Diving In

Now it's time to get working. I almost never begin at the beginning. Usually I examine the subcategorized list and I look for the toughest nut to crack. The real asskicker of a problem. The one for which I have the most difficulty imagining a solution at first glance. Once I find it—in the case of the ray gun it would probably be fabrication of the top scope—that's where I start. I do this for three reasons: 1) I don't want to get caught out toward the end of a project with unexpected problem solving that takes way longer than I expected; 2) once I've cracked the tough problem, I've built a lot of momentum, and I've already slayed the beast that might kill my momentum later on; and 3) I like coasting to the finish with the easy stuff. It's one of the ways I manage the stress of a project. Get the hard stuff out of the way first, then the specter of all those empty checkboxes becomes less intimidating because the tasks get successively easier and the checkboxes get filled in just as quickly.

Step 5: Make More Lists

Once you've made your proper nested master list, and you've dived into the making process, you're not done with list making. You're just at the beginning. As you go, new problems will arise, and new lists will become necessary to tackle their solutions. Additions, subtractions, and details you'd never have considered will rear their many heads. How do you tackle those? Yep, more and more lists.

I make lists most every day for each project I'm working on. As I move through the day building out my top scope, I keep crossing off each item in turn, and as the list of checkboxes starts to color in, I can see, in a physical, graphical form, how I'm doing.

There will always be things on the list that I can't get done that day, of course. There will be items that I can't start working on until other items have been completed. That is the nature of a complex build. Many parts are often dependent on other parts. A complete, detailed checklist for each component allows you to see those relationships clearly. It helps you to visualize the momentum of the project while acting as a bulwark against frustration.

Step 6: Put It Away for a Bit

All the list making I've described so far has been about project management and measurement and momentum. But there's one more list I regularly make, that stands apart from those. It's one I reserve for when I am not near a shop or any of my regular tools—an all-too-common occurrence these days. I often make this kind of list on planes, and in green rooms, or coffee shops. It's the list I make when I quit a project.

I frequently abandon projects, for various reasons. Life, travel, TV shows, more important projects—there are lots of reasons that I'll shelve some projects. Not permanently, but they'll lie dormant for periods of anywhere from days to sometimes years. In those cases, I find it handy to make a list of exactly where I stand with that particular build: what I've gotten done; what I was planning on doing next; what is needed for the next steps. Checkboxes here are vitally important, particularly for when I'm ready to come back to a project. I want to feel like I've made meaningful headway before taking a second crack at something, and seeing all those colored-in checkboxes really helps.

That may be the greatest attribute of checkboxes and list making, in fact, because there are going to be easy projects and hard projects. With every project, there are going to be easy days and hard days. Every day, there are going to be problems that seem to solve themselves and problems that kick you down the stairs and take your lunch money. Progressing as a maker means always pushing yourself through those momentum killers. A well-made list can be the wedge you need to get the ball rolling, and checkboxes are the footholds that give you the traction you need to keep pushing that ball, and to build momentum toward the finish.

USE MORE COOLING FLUID

My sin, as a builder, has always been one of impatience.

I want to get the part I'm working on done fast so I can move on to the next and the next. I want to take *all* the shortcuts to get to the finish line—to the finished product—as fast as possible. I'm *always* working against this impulse. I am the empirical test that demonstrates that building something with haste can take twice as long as building it with planning and forethought. A stitch in time really does save nine. And I've had a lot of stitches. More than sixty in total, 90 percent of which were the product of my impatience, I'm sure.

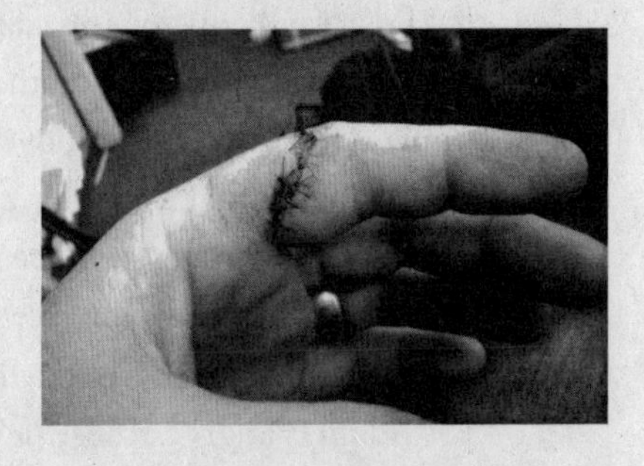

I've been this way my entire life. Growing up, my family went to Cape Cod every summer, to this little area where all the houses around us were owned by relatives. My great-uncle Paul Sheldon had a woodshop almost next door to our house. I spent many summers in that shop learning the rudiments of making small projects and also about addressing your work: clamping your work down tightly to better cut it with a coping saw; the basics of using a drill press, carefully marking your work before you bring down the bit to meet the work surface. We made shelves, we made marionettes. It was my first real maker space. I was supremely lucky to have had access to that little crucible of creativity, even if I could never be bothered to slow down and take the time to appreciate all of it.

Around age ten, I was in my Uncle Paul's shop, making a duck marionette he'd found the plans for in a craft magazine, and the next step in the process called for drilling a hole in the feet to thread a length of rope through to use as a tether. The marionette was composed of some simple feet joined by sailing line to an oblong body, all attached by monofilament to a T-shaped frame made of paint-mixing sticks. Uncle Paul taught me to mark every spot where you intend to cut or drill, but looking at the plans I was pretty confident that I could eyeball the drill hole. It was just a little hole in a foot, why would I need to actually mark the spot where I was going to drill for something simple like that? So I let it rip. When I was done, I blew away the sawdust and discovered that I'd actually missed my mark by a noticeable margin. It was a visible mistake—not just aesthetically displeasing but also structurally detrimental. The foot now hung at what looked like an injured angle, and I would have to use the laborious coping saw to cut a whole other foot, lest I be saddled with a lame duck.

I was so pissed. I got even more upset when someone came through the shop and asked why I seemed like a sourpuss.

"Adam is angry because he's too impatient to take the time to mark a hole he wants to drill," Uncle Paul replied, not even looking at me. I hated being pinned to the wall like that. I already had that tinkerer's sensibility of wanting to be an original, to be inventive and mysterious. But Uncle Paul saw what I did and was 100 percent right about me.

Even as I started getting serious about making stuff, thanks to a great art teacher in high school named Mr. Benton, impatience continued to dog me. Mr. Benton taught me a crazy amount of cool things from vacuum forming, to making custom chocolate bars, to sculpting in clay, and also how to use an airbrush. His open appreciation of my inquisitiveness was thrilling, and I kept adding new skills to my toolbox as a result. But having a voracious appetite for new techniques, and this being high school in the early 1980s before the internet, it wasn't long before I reached the limits of Mr. Benton's capacity to expand my making horizons.

So around the age of sixteen, tired of waiting for Mr. Benton to learn new things so he could teach them to me, I started spending more time at the library (you know, that old building with all the free books in it?), researching things I was curious about. Some of them were a little odd, or just arcane, like how to build sailboats (I was reading a lot of seafaring novels at the time), while others were more straightforward and foundational to the art of the build.

The one I remember clearest was reading an old military manual about how to tap and thread holes in metal. Living in this current era of IKEA, when you can assemble an entire house worth of furniture with a single tiny Allen wrench and an index card–sized instruction manual that uses no words and doesn't actually tell you what anything is, it's weird to imagine a time when you would ever need to look up something like how to drill a hole or even have to consider the need for a power tool to put something together.

The reality is, though, that both of these things—the technique *and* the tool—are fundamental to making. And when you are young, they are absolutely essential, because the sandbox of the budding builder is not *making* as much as it is modification:

taking something that exists and making it better, either functionally or aesthetically or both. Often that involves attaching and securing parts that were not originally intended to go together. That is where the hole tapping and hole threading come into play. You can imagine, with my level of impatience, how many modifications I was bursting at the seams to make to practically everything I owned. It was an itch *begging* to be scratched, and there was only one tool I knew of that could reach it.

THERE'S HOLES WITH THEM THAR DRILLS

When most kids turned sixteen back then, they wanted to get their driver's license and, if they were lucky, a shiny, brand-new car of their own. I didn't care about a car so much, but I definitely wanted something shiny and brand new—something that had *just* come out: the Makita cordless drill. Others might have come before it, but as far as I was concerned, this was pretty much the first cordless drill worth its weight. I told my parents that was what I wanted, and on the afternoon of my sixteenth birthday, my dad took me to the hardware store and bought me what would be the first of a lifetime's worth of cordless tools.

I used the spit out of that Makita for almost twenty years. What pure magic it was to wield such a thing. From a pug-nosed heavy corded drill—like the one I used to deform my duck marionette—to a motor untethered to a wall outlet! What alchemy yielded so much power inside of the battery to make it powerful enough for all my modifying needs?! It was utterly liberating. It also engendered some new bad habits, along with inflaming some old ones. After all, is there a better (or worse) tool for the chronically impatient and perpetually forward moving? I was about to find out.

Shortly after getting the drill, I was attaching a new rack I'd bought at a garage sale to my old bicycle. They weren't quite compatible, so I was making small modifications to get the rack and the bike to play nicely. The last step of the modification should have been simple. All I needed to do was drill a small hole into the structural arm of the rack so I could then secure it to the seat stay with a small bolt. (Me and drilling small holes, I swear!) The problem was, I'd already attached part of the rack to the seat post, so instead of utilizing proper leverage and drilling down into the metal arm against a flat workbench, I'd decided to drill up into the arm as it hung in the air, suspended from the seat post in open three-dimensional space. Imagine trying to core an apple with a knife while cupping the apple in your hand instead of holding it down on a cutting board. You can do it, but it's likely to get messy, and I'm not even talking about all the blood from when the knife slips and you gouge yourself in the palm.

Now, I could have taken the rack back off the seat post, clamped it securely to my workbench, and drilled the hole the "right way," but I was in a bit of a bind in that moment. I was running late to my summer job bussing tables, and I needed my bike to get to work, but I couldn't ride it without the rack being fully secured to the frame. I was in a hurry. Plus, if I'm being honest, I was feeling pretty cocky with this new Makita in my hands. I figured I could just lie on the floor of my workshop (aka my parents' basement), use one hand to hold on to the rear wheel of the bike to keep it steady standing up, and use my other hand to quickly rip the hole I needed. Sure, I wasn't in the most ideal position (I was actually in the *least* ideal position), and yes, maybe I was drilling upward from the floor and way off-angle, but what's the point of a cordless drill with all this power if not for tasks precisely like this one?

There are many points to a cordless power drill, it turns out, and the most important is the one at the end of the drill bit you're using to make the hole you want. I did not fully appreciate this fact when I was an impatient sixteen-year-old, but I was about to find out the consequences of taking it for granted. Lying in a weird position, using only one hand for support, made the bike shaky. This meant I couldn't properly feel the resistance I was getting as I hastily and haphazardly pushed the drill bit through the metal, which meant, in turn, that if the drill bit started to get too hot, I wouldn't notice until it was too late.

Too late for what? To answer that, we need to talk some real physics.

COOL IT NOW!

Cutting something is about the relationship between the thing you're cutting and the thing you're cutting *with.* Pretty much always, the thing you're cutting with needs to be harder than the thing you're cutting. In my case, both the structural arm of the rack I was drilling into and the drill bit I was using were made of steel (I've literally never encountered any other kind of drill bit), but for the process to work properly, the drill bit had to be made of harder steel than the rack.

How do you get steel to be harder? Great question. The answer is structure. The internal structure of the atoms and molecules that make up the harder steel has been "adjusted." One of the most common ways to do that is with heat. Heat can do magical and amazing things to steel. You can very finely adjust its ductility (malleability), its flexibility, or its hardness, for example, solely with the process by which you heat the steel and how quickly you let it cool.

Heat can also be your worst enemy, however. In addition to its hardness, the reason a drill bit can cut through steel is because it has a precisely ground cutting edge. As that cutting edge revolves, it is angled just right, so that it peels up a small shaving of steel from the work with each revolution, all while the long spiral flute of the drill bit pulls the resulting shavings up and out of the hole to keep everything clean. This process creates a tremendous amount of friction, of course, and with friction comes heat. Friction is the reason your car engine heats up, it's the reason your computer gets warm (the friction caused by all those electrons dancing to make your Photoshop filters work quickly—I'm serious), and it's the reason a drill bit heats up. Advancements in drill bit hardness and sharpness have been made over the years specifically to reduce friction as much as possible, but those efforts can only go so far.

I said you can do amazing things to steel with heat, but you can also do amazing damage with it. If you don't do everything (or enough things) right, you risk generating too much heat from the friction between the drill bit and the work surface. Because of this heat, the work surface can then become as hard or harder than the steel of the drill bit, a process called "work hardening," whereby you are hardening the material by virtue of the work you're doing to it. When you bend a paper clip until it breaks, for instance, the bending you're doing creates friction between the atoms within the paper clip, generating heat until the metal changes from a ductile steel to a brittle steel, and then breaks. For drilling, unintentional hardening is bad. Very, very bad. As the work gets harder, the drill bit's efficiency decreases, the friction *increases*, and then so does the heat, creating a destructive feedback loop. (In my experience, the sound of the drilling also changes.) Fortunately, there's a simple solution to this problem: COOLING FLUID.

Cooling fluid isn't a specific product, it's literally any liquid that helps move heat. It could be water, though usually it isn't because water is an efficient oxidizer of many metals (it's a perfect coolant for acrylic), but it is most always a fluid since liquid has far better thermal conductivity than air. Ever notice that 70 degree air feels warmer than 70 degree water? That's thermal conductivity in action—the water *feels* colder than the air simply because it's better at moving heat. Same goes for drilling or cutting. Cooling fluid keeps the work and the drill bit's cutting edge at a stable operating temperature. When using a metal-cutting saw blade in a band saw, the addition of cooling fluid visibly changes the speed of the cut, as it helps move both heat and debris away from the area being cut. If you're holding on to a portable band saw (one of my favorite tools in the world) you can *feel* it cut faster when you start to add a coolant to the mix. I love that feeling.

Of course, I didn't know any of this when I was sixteen years old. This is all stuff I realized later, after years of drilling and cutting everything under the sun: glass, rubber, cloth, leather, plastic, cord, rope, string, wire, aluminum, zinc, steel, and even titanium. I've cut cars in half. I've cut into bowling balls, airframes, computers, bicycles, and tires. I've used saws, disks, lasers, blades, chisels, wedges, and even plasma. I've cut things unintentionally, like myself (a lot), but inevitably the cut I learned the most from was this one in my parents' basement that ended up not being a cut at all. As I pushed the spinning drill bit hard into the rack's structural arm, the drill bit got hot, then it got dull, then I pushed harder to make up the difference, until eventually the bit seized and snapped, leaving a chunk of hot, work-hardened steel stuck forever EXACTLY where I really needed a hole instead of a hardened steel plug. With this one dumb, impatient movement, I'd achieved the exact opposite of my intention: I'd ruined the rack, destroyed the drill bit, AND made myself late for work.

I often say that if I could go back to any point in my past and tell my younger self one thing, I would go to this moment and say: "USE MORE COOLING FLUID." I know that sounds like a pretty trivial use for a time-traveling adventure—Marty McFly went back to save Doc Brown's *life*, after all—but for *most* processes involving the cutting of *most* metals, cooling fluid aids and abets a precise, predictable, and repeatable cut. It helps put holes where you want them. It prolongs the life of blades and bits, and it prevents tool failure. Before I knew this secret, I broke a lot of bits, destroyed many otherwise perfect parts, and cursed both myself and my tools in countless creative ways.

Beyond this simple good shop practice, though, the phrase has taken on a deeper and broader meaning for me over the years. Using more cooling fluid is a reminder to myself to *slooooow doooooown*, to reduce the friction in my life—in my work, in my schedule, in relationships, everywhere, really. It's a warning against my proclivity for serial impatience.

More than anything, though, "Use more cooling fluid" is an admonition to do all the things necessary to properly *address* my work.

ADDRESSING YOUR WORK

If impatience is my biggest sin as a maker, its primary manifestation has been in my continual struggle with addressing my work. In this context, addressing your work means orienting yourself, physically and mentally, in the optimal position to execute the task, despite the fact that it might take a little longer. It means taking the time you need to do the job right the first time. Taking the time to organize your thoughts; to organize your work space; to organize your tools. It's time, when taken, that you might feel is slowing you down in the moment, but in fact is saving you time in the long run.

I know this is true . . . in my brain. And yet, I've wrecked things countless times since Uncle Paul called me out at ten years

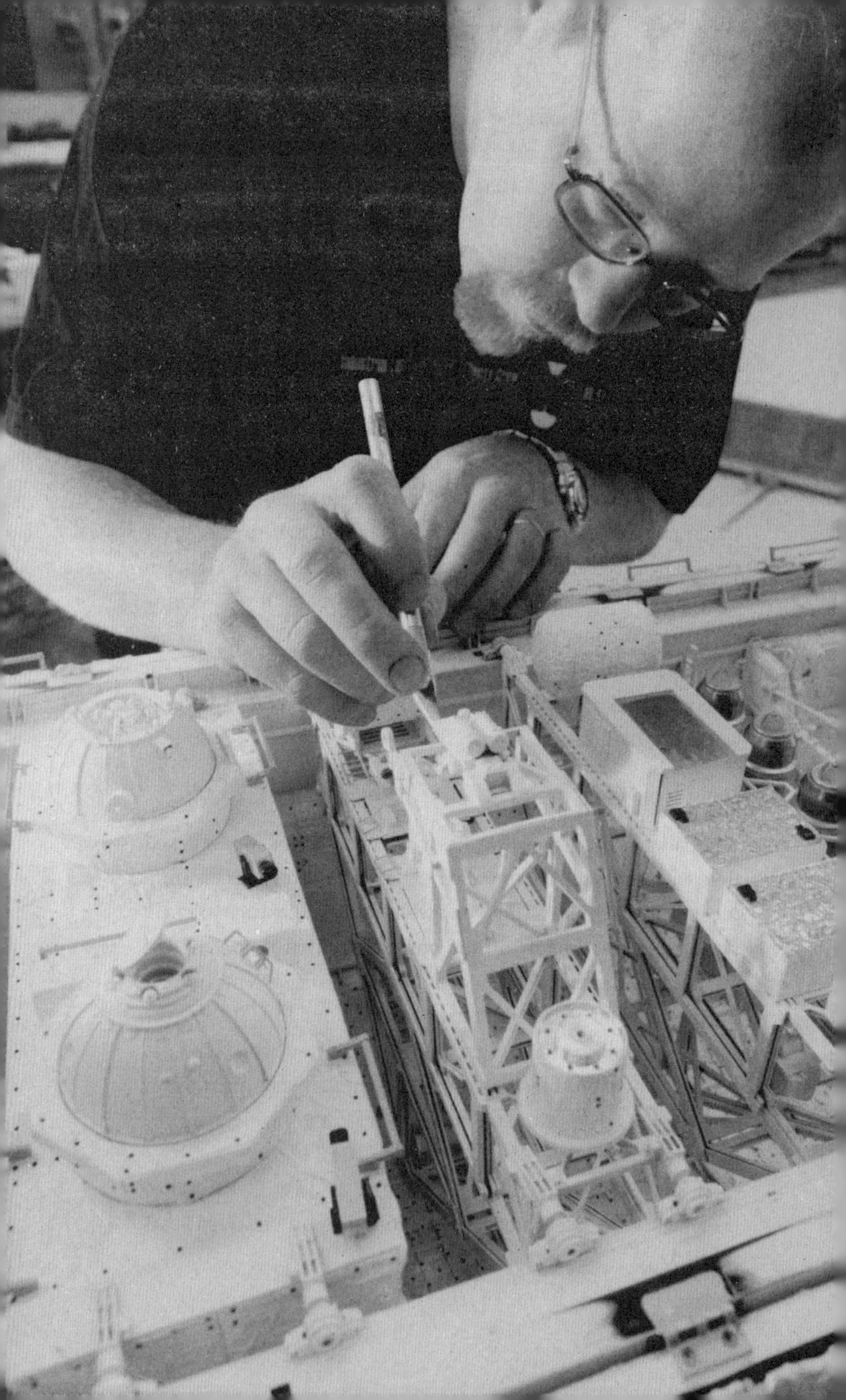

old. I'm in my fifties and still, in my natural state, I am driven by the desire to FINISH. Damn the torpedoes and damn the consequences.

To be fair to myself, impatience does have its upsides. Deadlines and goals are very useful tools. They refine decision trees, and provide momentum when things get boring. Whenever I'm doing something tedious, I'll admit to myself that it can be crushingly boring to do the same thing time after time for hours, and then I will use that bit of my own psychology in a constructive way and set arbitrary goals for myself. If I'm making something like twenty small brackets for a space shuttle bay gantry (as I was for *Space Cowboys*), I'll quickly throw down the gauntlet:

"I want to finish machining this part before it's time to head to dinner."

"I think I can make all twenty of these before lunch."

"It'll be soooo cool if I can see this section finished today, and ready for paint!"

As I'm doing work like this I have found that I can really dive headfirst into the meditative aspect of the tedium itself. Just one action repeated after the other: the glory of monotony. As I go I'm not zoned out, I'm constantly doing two things: first, I'm counting how many I've done in how much time, and then I'm doing mental math to see if I will be able to finish in some arbitrary time I've established in my head. The second thing I'm doing is asking myself if I could do it faster. The answer to that question will always be "no" unless I have first asked a number of other important questions related to how I am addressing my work.

What is the scope of my work space? What materials am I working with and how much of them do I have? What tools do I need? Do I have them all? Are they in a good place? Are they in *the best* place? Could I put the glue cup closer? Will that save me time? Is there a custom holder that I could assemble that would allow me to paint these things faster? Maybe a multilevel drying rack will save me a few trips to the paint booth. Balanced efficiency is one of my coping mechanisms for tedium. I'm on a constant hunt for refinements in my assembly process. In addressing my work, I love that kind of refinement. I love looking at a pile of parts that had daunted me two hours previously, but now sits finished and conquered. My friend Tom Sachs, an amazing contemporary artist from New York with whom I share a deep fascination with space, would argue that there should never have been a "pile" of parts on my workbench to begin with.

TOM SACHS'S TEN BULLETS

1. Work to Code (work within the system)
2. Sacred Space (the studio is sacred)
3. Be on Time
4. Be Thorough
5. I Understand (give/get feedback)
6. Sent Does Not Mean Received (get confirmation)
7. Keep a List
8. Always Be Knolling
9. Sacrifice to Leatherface (take responsibility for mistakes)
10. Persistence

Tom has ten primary rules for his shop. He calls them his "Ten Bullets" (he even made a short film about them) and they're all great, though the one I love the most is Bullet Eight: "Always Be Knolling." Knolling is an organizational process that Tom learned from, of all people, the janitor at Frank Gehry's furniture shop in Los Angeles when he worked there in the late 1980s. Every day, the janitor—a fellow named Andrew Cromwell—would come into the shop to sweep and vacuum. But first, before he got to any of that, he would go to each workstation and neatly line up all the tools and materials that were still out on the work surface in parallel lines or at 90-degree angles to each other. One day, Tom was still in the shop when Andrew came in to set about his normal routine, and he was amazed and delighted by what he saw.

"Andrew, what is that?" Tom asked. "That's so beautiful, the way you line everything up, what is that called?"

Andrew shrugged. He looked around, looked up at the wall, saw a sign hanging up for the company that was sponsoring the project Tom and his team were working on (the furniture designer, Knoll), and said, "Knolling?"

"So it just became that," Tom told me. "And I picked it up and started using it in my everyday life."

"What was the original point of it, though?" I asked.

"To make his job easier, I guess."

Boy does it ever.

KNOLLIN' WITH MY HOMIES

When I first met Tom I'd been knolling for years but had no idea that what I was doing actually had a name. I knoll my workbenches, I knoll my desk at home. When I was fourteen years old, I bought ten pounds of locks and keys from a flea market, then dumped them out on the floor of my room and proceeded to stack them by lock type and category. I knolled them. Today, every time I check into a hotel room I dump my bag out and, as I check in with my wife on the phone, I slowly knoll the contents of my bag. Sure, it might be ever-so-slightly OCD in its execution, but knolling is a great way to take stock of what is in front of you. I don't lose things in my bag, either. At any given point my daily bag has over a hundred things in it, and I know what each one is and where it is. This gives me such a feeling of relief.

Here's how to do it, according to Tom (and, really, common sense):

1. Examine your work space for all items not in use—tools, materials, books, coffee cups, it doesn't matter what it is.
2. Remove those unused items from your space. When in doubt, leave it on the table.
3. Group all like items—pens with pencils, washers with O-rings, nuts with bolts, etc.
4. Align (parallel) or square (90-degree angle) all objects within each group to each other and then to the surface upon which they sit.

For a large part of my career, I worked in a consistent fashion: on an eighteen- by eighteen-inch square, in the middle of an overcrowded bench, surrounded by piles of stuff. My hoarding tendencies and my general impatience made any other kind of work environment a virtual impossibility. Knolling was an epiphany on the same level as adding checkboxes to my lists. Beyond giving my brain the space to more easily take in what I am working with, knolling my work space throughout the day also reduces the likelihood that I'll lose things, and increases the likelihood that I will find them quickly when they do go missing. It also creates a lot more space to work in as a result. By forcing me to slow down (which I've learned actually allows me to work faster), the whole process saves me time on the other end of my work process. For Tom Sachs, who has been working out of various New York City studios since the early 1990s, the value was more practical. "It was the product of living and working in a high-density environment," he explained.

OPPOSITE: This travel bag was pretty well knolled, though the cords and razor components could have been better. And yes, I'm terrified of running out of hearing aid batteries.

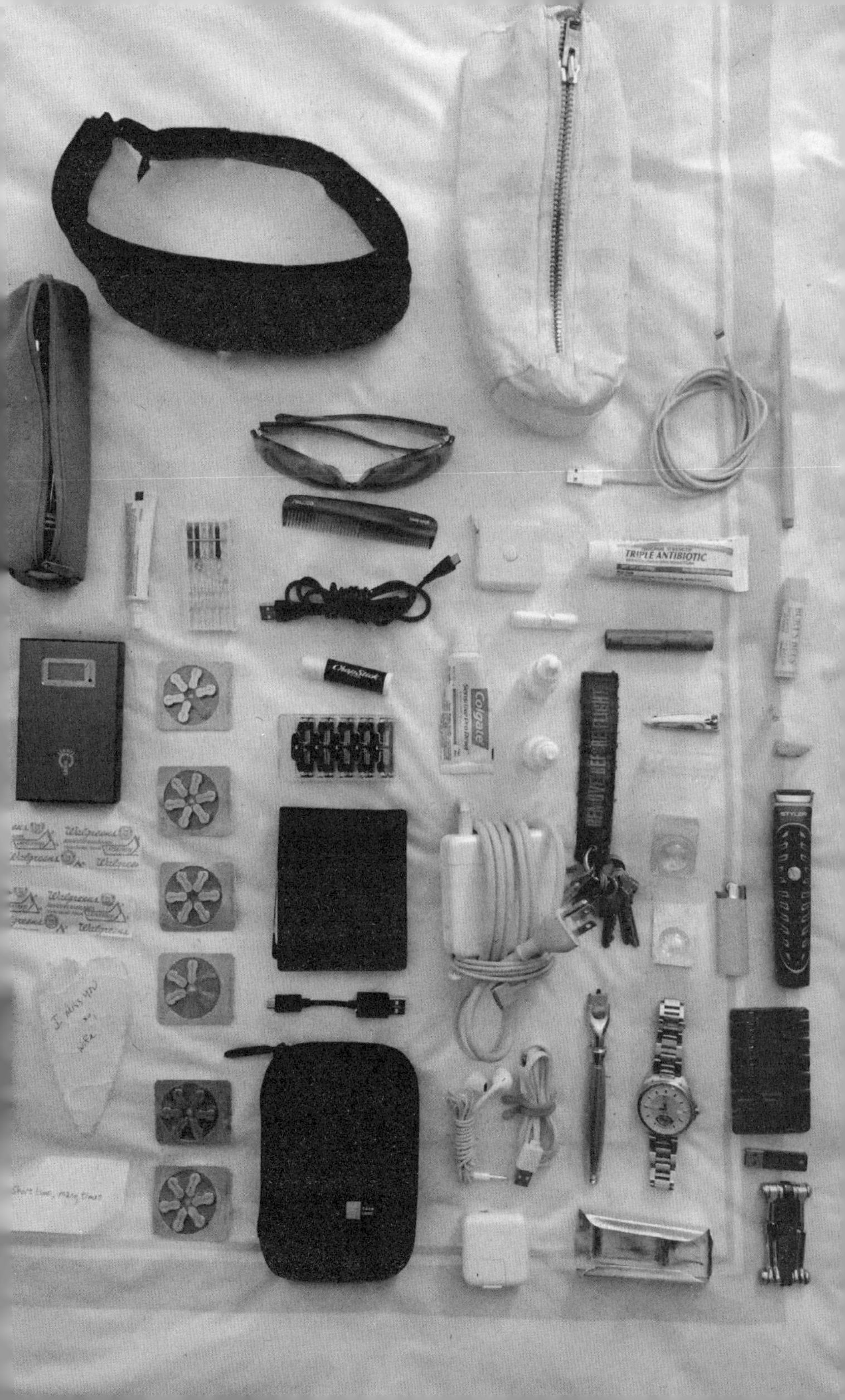

TRIPLE ANTIBIOTIC
ChapStick
Colgate
REMOVE BEFORE FLIGHT
Walgreens

There's one group of makers who understands intuitively, perhaps more than anyone else, the multifaceted value proposition that knolling represents: chefs. They call it *mise en place*. Coined in the late nineteenth century by famed French chef Auguste Escoffier, *mise en place* translates roughly to "everything in its place." The principle behind it, cribbed from Escoffier's military service during the Franco-Prussian War, is really about order and discipline. James Beard Award–winning chef of San Francisco's Jardinière (and my son's boss), Traci Des Jardins, describes *mise en place* more plainly: "It's so inherent to what we do, it's just what we call having your shit together."

Like Tom, Traci is no stranger to working in high-density environments. That phrase pretty much describes every busy restaurant kitchen during service, and Traci has spent years in some of the busiest and most famous: from the world-renowned, three-Michelin-star La Maison Troisgros outside Lyon, France, to Patina in Los Angeles, to Michael Mina's early seafood standout Aqua, in San Francisco, to now a half-dozen restaurants of her own.

In a busy kitchen, "you have to go through a lot of processes before you get to service, which is when you're actually putting meals together on plates," Traci described for me. "*Mise en place* refers to all of these various components that are prepared ahead of time and you have in place for that moment when you get the order and you put the dish together."

Kitchens are pressure cookers in which wasted movement and hasty technique can ruin a dish, slice an artery, burn a hand, land you in the weeds, and ultimately kill a restaurant. *Mise en place* is the only way to reliably create a perfect dish, to exact specifications, over and over again, night after night, for paying customers who demand nothing less.

Nowhere in the cooking world is this more true than with baking. If cooking is a passionate, creative, artistic pursuit, then baking is an exacting scientific discipline. And surprise, surprise, Traci Des Jardins is a terrible baker, by her own admission. "I hate measuring things," she said when we talked about her attempt to make banana bread over at her partner Jennifer's house one day.

"I usually resist *mise en place* in baking, because it kind of feels like it takes up a lot of space, and I think, 'Do I really want to dirty all those bowls? Can't I just put the flour on top of the eggs?' But in this case, I was scaling up a recipe three times, which is a recipe for disaster if you're just doing all that math in your head—you can really screw it up—so I approached it a little bit differently than I normally do. I pulled all the bowls out, I put all my dry ingredients in one set, I had all my wet ingredients separated in another, and then I put it all together and it was so much easier to do it that way. Turns out it works a lot better to kind of spread it out and have everything in its place and then put it together."

The stakes are not dissimilar for a maker in their workshop. For all the alchemy that goes into building something, the magic of making is only possible because of the many repetitive processes we endure in preparation for final assembly, and then the deliberate way we put it all together. And the only way to get that prep right, to get your *mise* in its proper *place* is to slow down, address your work properly, and lock your shit down. In the case of a maker, I mean that literally.

WELCOME TO THE FUTURE CLAMP-ITS

If I showed you a picture of me cutting a piece of wood, next to a picture in a woodworking magazine of a dude cutting the exact same piece of wood, do you know what the difference would have been as recently as just a few years ago? I would have been using my foot to hold the wood down across the middle shelf of one of my rolling carts (with wheels that may or may not be locked), while the dude in the picture would have the wood clamped so securely in a vise at perfect working height that he'd only *need* one hand to operate his saw (though, of course, he'd be using two) and he could use the other to drink a piña colada or steal my high school girlfriend. Jerk.

To me, in this sense, woodworking magazines are clamping porn. And they exist for a good reason: precise work requires precisely acting *upon* your work, and good role models to show you what a proper setup looks like. I'm sure to some of you that sounds like common sense, and I envy your patience and wisdom. But right now I'm talking to the group of aspiring makers out there who don't automatically know that it's unsafe to use a 1¼" Forstner bit on a piece of wood that isn't firmly clamped down. Or that holding the wood firmly in your off hand does not count as being "firmly clamped." (Yes, I've made this mistake, and yes, I've got a picture of the injury, and no, you don't want to see it.*)

Fortunately, we live in a good time for clamps. In the late '80s, IRWIN, makers of the incomparable Vise-Grip locking pliers, came to market with a quick clamp that was perfect for impatient people like me precisely because you can use it with one hand. (I have more than twenty of just this one type of clamp, and I use them all the time.) And it's not just clamp technology that we have to celebrate. There is a whole host of tools and materials that have made making safer, more effective, and more efficient. But it's about more than just clamping and cooling fluid, or even sav-

* Someday I hope to do a lecture on hand injuries sustained from stupid moves like this, but it will have to be at a medical school or someplace where the audience can handle all the images in my "injury" folder.

ing time. It's about looking forward, into the near and far future, and making an assessment of what you truly value, so you are not reckless with it, or so impatient that you don't do what's necessary to see it through to fruition.

I know that sounds like hyperbole, but it's not. Trust me, you've never tried to make a fourteen-foot floating balloon out of twenty-eight pounds of rolled lead. That's right, on one season of *MythBusters*, Jamie and I had to make a lead balloon, with lead so thin it had the material properties of wet toilet paper (not an exaggeration!). It would fall apart at the slightest provocation. If you crumpled it even a little bit, tiny microholes formed. If you crumpled the whole piece, it would look like a lace curtain if you ever managed to unfold it and then hold it up to the light.

The lead balloon build demanded I fight against every one of my natural impatient instincts. Just getting to the place where we could consider testing it took *two years* because no one could make lead thin enough for us.

Lead is formed into sheets for many different industries, but in general we found that it was rarely made thinner than a couple thousandths of an inch—roughly half the thickness of a human hair or a piece of paper. The problem was that at a couple thousandths of an inch thick, our balloon wouldn't float. We did the painstaking volumetric calculations for the size of our balloon in conjunction with the lifting power of helium, and it was clear that we needed our lead to be no more than one-thousandth of an inch thick. Two companies told us they could do it but broke their equipment trying. Finally, Jamie found a company in Germany that was able to do it even thinner. They made us several hundred square feet of lead that was .0007" thick.

Once we had the lead, the process of making the balloon itself (and making it work) took so much discussion and planning and

deliberate care that one of our producers even sought out the counsel of a local origami specialist. But that was a problem right from the get-go, because origami is about folding and we obviously couldn't fold this stuff (nor could we just grab it and throw it across the balloon's frame, which I secretly wanted to at points). We had to assemble it like it was stitched together from pages of the Gutenberg Bible (that is, carefully), then arrange it in such a way as to distribute the strain of being filled with helium across the greatest totality of the structure without making a single mistake. If we let the pressure build at any one point for an instant, the whole endeavor would come crashing down.

Eventually, with some time and a lot of deep breaths, I realized that we could create a quasi-spherical balloon by filling a cube with pressure. The cube, with its six square sides, could be broken down into a set of smaller triangles, which could likely be cut flat and then strategically moved in place and taped along each side in turn. It was the only formulation that seemed like it could work and involved, BY FAR, the least amount of movement of the material. And thank goodness it worked, because the only thing more fragile than the sheets of lead at that point were my nerves.

When we finished all the important filming, to demonstrate the project's fragility in real time, Jamie took down the balloon by throwing a baseball at it. Or more properly stated, *through* it. The baseball passed through the balloon like it wasn't even there. Then the entire balloon collapsed and came to the ground with a surprisingly loud *thud*. That's how close we were to catastrophic failure.

Keeping that balloon from failing while we constructed it required what Jamie referred to as "the closest we'd ever get to seeing the future." We were working with each other, running countless failure scenarios in our heads, and cutting each off at the pass before it had the chance to bone us. THIS is what I mean by addressing your work. It's not too much to say that in slowing yourself down, locking your process down, and doing everything the right way, you're looking into the future that you want to create, with the things you value at its center.

DO UNTO OTHERS

There will likely come a time in your life as a creator, if it hasn't happened already, when you take the leap from making things entirely for yourself to making things for other people. I don't mean as gifts. I'm talking about commissions. Jobs. *Work*. Things for people who have seen your earlier stuff and said, "Yes, I want something like that. How much?"

The first time it happens, it's equal parts thrilling and terrifying. The idea that someone is willing to *pay* you to do the thing you'd have paid them to let you do for free the day before is so incongruous to your amateur's brain that it almost feels like you're running a con. "Wait a minute! Who are you kidding?" your brain says to you, "You don't have enough experience for this, you're still that kid who couldn't even attach a rack to a bike seat . . ." This thought, which is effectively imposter syndrome, often couples with self-doubt, panic, fear, and uncertainty to tip the scales in the constant internal battle between the security of doing what you've always done, more or less consequence-free, and the exciting demands of doing something new for somebody else. The goal, if you are ever to grow and succeed as a maker, is not necessarily to win that fight every time but to find the balance between the two warring parties as you go about creating work. You don't want to lose who you are, but you also don't want to keep making silly mistakes that used to only cost you your time, and now could cost you a job or your reputation.

My first true experience making something in exchange for a commission was a great ride, but sadly overshadowed by what came next. In the mid-1980s while I was at the NYU Tisch School for the Arts I fell in with a group of students at NYU's famous film school, many of whom are still my great friends, who always needed help producing their low-budget student films. Well, technically, I wasn't *at* Tisch as much as I was *around* Tisch, having dropped out six months into my first year in the acting program. The first big project was my friend David Bourla's senior thesis film, an ambitious fantasy called "Gargoyle and Goblin" about a detective agency run by, you guessed it, a gargoyle and a goblin. (When you can get two awesome monsters *and* alliteration into your title, you know you're onto something.) David and I talked science fiction and fantasy films constantly. We were kindred spirits in our love for the unbelievable and barely imaginable. He brought in a small crew, including me, who had that same sensibility and could also build sets and make props. I was living in Brooklyn at the time, making sculpture from anything and everything I could drag off the street, and, being generally available with nothing else to do, I was an obvious and enthusiastic choice.

Shot over sixteen nights in an abandoned block of buildings in Hell's Kitchen, the film demanded a little of everything from our versatile crew's tool kit. Mike, our director of photography, probably had the strongest skill set at that point. He was our shooter, and he was also in charge of the animatronics for the gargoyle's wings. I had my hands full, too, with prop building. I cooked up massive batches of corn syrup blood for the vampires and headless corpses. My dad's antique glass bottle collection stood in as the vessels for a stockpile of potions in a wizard's office. I made cauldrons, coffins, and built a complete padded cell for less than one hundred dollars (and a lot of staples in foam). The lion's share of the work, however, involved building over a dozen unique sets that, with the intrepid art department to which I was assigned, we had to dress for every shot. It was a lot of work, but the work paid off. The ambitious film won several awards at NYU's student film festival and, on the back of that success, I was asked by another good friend, Gaby, to do the art direction and set construction for another senior thesis film she was producing.

Unlike David's alliterative adventure full of ghouls and gore, this film was a small comedy about a man with a terrible toupee trying to retrieve money from a wisecracking, intelligent ATM that subjects him to ever-increasing humiliations with the intention of forcing him to remove his ridiculous hairpiece before it would dispense his cash. After meeting with Gaby and the director, we realized that the whole film could actually take place in a single location—the ATM vestibule. Great for the locations' budget, not so great for the props' budget, because we couldn't just borrow an old ATM to use for a few days of filming. This was the mid-1980s. There were no old ATMs! Sure, the technology had been around for twenty years by this point, but ubiquitous, standalone ATM machines weren't a thing yet. You couldn't just walk into a gas station or a bodega and expect to find a cash machine. ATMs were big hulking beasts bolted into the sides of the bank branches that operated them. I'd need to build our ATM from scratch as well as the vestibule along with it.

The ATM build was an exciting challenge that ate a fair amount of the production budget, but one that, still flying high on the accolades from "Gargoyle and Goblin," seemed straightforward and doable. One location? A few walls? An ATM that would effectively be a large arcade game cabinet retrofitted with a keyboard and some buttons? No sweat. Gaby found a location at a friend's apartment in Brooklyn where I could spend a month building the set. That first day, thrilled beyond belief to be working on another cool film for another great friend, was pretty much my last good day on the project.

WHAT DO I DO?

Things started going wrong from the very beginning. In the first week, I built and painted the flats (that's theater talk for "fake walls") for the vestibule, using wood frames and canvas like I'd seen set crews do in high school theater. Unfortunately, I didn't realize you had to stabilize canvas flats before painting them, otherwise they tend to wrinkle as the paint dries. When I showed up to set the week before the shoot to erect the vestibule walls, all of my flats looked like satellite photos of the Utah desert—bumps and crags and fissures and creases all over the place. I spent the next three days straightening them (unsuccessfully). Those were the three days I had slotted for construction of the ATM shroud. I had foolishly saved the set build for the end because I thought it would be the most fun to work on. Of course, just like the flats, the ATM ended up kicking my butt. At one point, I accidentally glued the piece of acrylic I'd carefully cut to surround the screen of the ATM to its backing, and cracked it in half trying to peel it back up. There was nothing to do but secure it back in place, with the crack in plain sight, because I had not a penny left to buy another piece of acrylic. Mistakes piled up like that, and their consequences became more and more obvious.

What is still the most insane part of all this was that it didn't even occur to me until about five days before shooting began that I was in over my head. Why would it? This was film! Gritty, indie "cinema of transgression," they called it at the time. Plus, I was young, I was a wunderkind! I was going to make this work against all odds. No matter how far behind I got, I figured I could always forgo a little sleep, throw some of my boundless manic energy at the tasks in front of me, and then they'd just kind of sort themselves out. This was not just another mistake, it was a full-on delusion (though it seemed like a perfectly rational plan to me at the time). There is no skill in the world, I have since discovered, at which you get better the less sleep you have.

Filming was slated for three days over a long weekend. When the crew showed up on location ready to go Friday morning, they expected to find a completed set, ready for filming. Instead, they found me, frazzled after sixty hours without sleep, standing in the middle of a set in which every single last little part had something wrong with it. The most unforgivable was the cracked and unusable ATM shroud. The film featured many tight shots of the ATM, and mine was not ready for its close-up. The sign above it wasn't hung straight, but the holes had already been drilled and couldn't be hidden again. The walls were still wrinkled. The doors didn't open. The linoleum tile floor I had laid down over the carpet in the apartment was buckling, and I didn't have a plan to fix it because, like everything else on this production, I had never done that kind of thing before. I had zero institutional knowledge.

Every film crew, as David Mamet says, is an army of problem solvers, and this crew was no exception. They gamely rolled up their sleeves and started helping with everything they could. They kept asking me what they could do, but having never truly delegated before, I had no earthly idea. I spent the entirety of my time—*a month*—building this set thinking that I had everything under control and now that I didn't, now that I clearly needed help, I had no idea how to utilize it.

After a few hours of this farce, one of the crew members turned to me and asked directly, and angrily, "Do you even know what you're doing?"

I remember his frustration so clearly. What do you say to that? How do you respond to a question that hangs over a room and challenges your whole identity? I'll tell you how I responded: as Indiana Jones. A situation with this much tension could use some levity, I thought, so in my best distracted Harrison Ford voice, I replied, "I don't know. I'm making this up as I go along." My joke did *not* go over well. Not because it wasn't funny, but because it wasn't a joke. The crew member put his hand on my shoulder, looked me dead in the eye, and said, "Go home."

Being sent home was humiliating. Yet I was grateful to be released from that particular hell. I left my tools right where they were, walked home, and did not come back to retrieve them until well after the shoot was wrapped and everything else had been loaded out. When I returned I found a note from Gaby: "CALL ME." If you are conflict averse, like I am, this is the kind of note that burrows a hole in your gut and pumps blood through your heart with such force that you can feel it in your throat and hear it in your ears. I rang Gaby when I got home with my tools. She spoke calmly, but as she walked me through the laundry list of my transgressions it was crystal clear she was enraged. The director worked for three summers to save enough money to make his senior thesis film and I'd basically ruined it. The crew had to pull three all-nighters to fix *my* problems. She said, and I will remember this for the rest of my life, that I couldn't have done anything worse to make it clearer to her that she should not be my friend. It was like a reverse Hallmark card.

I have never felt lower.

I called my dad. I needed some guidance. Some kind of advice. I didn't have the language or the emotional awareness for it at the time, but what I really needed was help. *What do I do?* My dad told me there was nothing he, nor I, could do to make me feel better. This situation sucked no matter which way you sliced it. My only choice was to recognize that I'd messed up, and realize that while I'd made a series of dumb and self-inflicted mistakes, this didn't define me as a bad person. He told me to log the experience and examine why it happened, so that I didn't make those same mistakes again when another opportunity came around.

In retrospect, the most obvious mistake I made was thinking I could build the whole set by myself. At the time, this mistake did not seem so self-evident. I was always pretty clever at figuring stuff out on my own. Whether it was circus arts or solving Rubik's Cubes or putting things together, I'd become obsessed with a skill, then study and learn and practice it until I was better than most people. But "better than most" at casual stuff does not translate to "good enough" when you're being paid to bring someone's vision to life. You need all hands on deck when you're hired to make something for somebody else, and at nineteen or twenty years old, mine were the only two hands whose use I knew how to recruit.

TRYING TO BE A HERO IS A TERRIFIC WAY TO END UP BECOMING THE VILLAIN

Even if I had all the necessary skills and experience to execute a job perfectly, going it alone without any help would have been foolish. Not only is it less efficient, but how do you expect to learn new things or get better if you do everything in isolation? This was my biggest mistake, my truest and deepest failing: my aversion to asking for help. For more years than I would like to admit, "help" was a dirty word. I was great at giving it, and I never judged anyone who needed it, but I was terrible at asking for it, because it felt like failure—a very specific kind of failure that was unique to me.

Realizing I had an aptitude for making felt very much like the epiphany at the center of a superhero's origin story: I held immense power but I had no idea how to corral it. I had no sense for what its limitations and potentialities were. All I knew was that I didn't want to give it up, I wanted to use it for good instead of evil, and that it was entirely up to me to figure out how to do that and get better in the process. My strategy back then might sound familiar: I'd accept a new challenge, then go off alone, put my head down, work hard, work fast, try stuff, and see what happened. It was my impatience rearing its head again. And, help? Well, that's something heroes give, not something they ask for. And I wanted, more than anything, to be the hero.

Let me tell you something: when you are working in a team, making something for somebody else, trying to be the hero is a terrific way to end up becoming the villain. As a collector of skills and arcane trivia, I get a huge endorphin rush from knowing the right answer to a question, or having the right tool to fix a problem. But I've had to learn (the hard way, in some instances) to be really honest about what I do and don't know. It's bad enough when you try to fool others into believing that you know what you're doing; there's zero percentage in fooling *yourself* as well, because that only puts you further behind. I had to lose a friend, which still makes me sad, to learn this lesson. I had to ruin a fellow creator's dream in order to understand that having the patience and humility to ask for help is a critical element in the successful execution of any project.

The most surprising aspect of learning that lesson, as it burrowed itself into my brain very much against my will, was that when trying to make something work, it was always the smartest people I knew who were the quickest to ask what the hell you were talking about; to ask you to explain; to ask you to help *them* understand. In this sense, help is more than just an extra pair of hands or an extra set of eyes. Help is expertise. It's wisdom. It's learning enough to understand and admit to what you *don't* know. It's how you learn to do new things and how you deepen your skill set.

It is also, fundamentally about collaboration, whether you are an apprentice, a partner, a peer, or a boss. After kicking around Brooklyn for a year as an artist, helping my friends produce their student films, I put a bunch of energy into trying to get work at a series of small FX houses in Manhattan. Sadly, none of those places was an environment in which I was encouraged to learn anything and, unsurprisingly, they were unpleasant places to work. I gave up on them for stable work in graphic design and then moved back home for a year to try to figure out what my next steps would be.

My parents were more than safe harbor, more than a place of last resort, however. I enjoyed the privilege of choosing where I was going to work, and who with, at such an early stage in my career because they covered many months of rent for me when I was between jobs. This is a luxury not many have, and one for which I'm extremely grateful. As a parent, I can now see they were investing in me, but there were plenty of times it must have felt to my parents like an investment of questionable merit.

My problem wasn't *really* those gruff effects houses, after all, it was that I didn't actually know what I wanted to do. I didn't have a specific ambition, and the fact is New York City is a hard place to survive if you don't know what you want to do. I needed a little elbow room to figure it out, so in 1990 I packed up my meager belongings and moved three thousand miles west to San Francisco where I stumbled fortuitously into the city's red-hot theater industry.

In San Francisco, I quickly secured a job as assistant stage manager for George Coates Performance Works, an experimental theater company that was a pioneer in pushing the envelope of multimedia live performance. Their productions featured 3-D projections, large stage illusions, early Computer Graphics computers, and large, elaborate mechanical props.

At Coates, and other theaters around the city during those years, I absorbed a wide range of skills, largely by doing the opposite of what I'd done in New York. Instead of accepting anything that seemed exciting and then going off on my own, pretending that I knew what I was doing, I asked for help when necessary, and gave help whenever wanted. In the process, I learned carpentry, set design, casting and mold making, rigging, costumes, furniture construction, and welding. I spent weeks learning scenic painting from a master of the form, Shevra Tait, who normally worked for the San Francisco Opera House. Needless to say, there were no wrinkled flats on a Shevra Tait set. Patience and diligence and a dry sense of nihilist humor were the order of her days. The list of skills I learned by asking for help is long and I loved every minute of it. Most importantly, after dozens of all-nighters getting massive shows off the ground across the Bay Area, I learned the singular pleasure of busting my ass as part of a creative crew where everyone was rowing in the same direction.

Eventually, my work on the theater circuit was brought to Jamie Hyneman's attention. He ran an effects house for a film company called Colossal Pictures that did a ton of commercial work. Jamie was always looking for people who could work under pressure, on a wide array of jobs, and learn quickly. After an hour-long interview to which I brought a suitcase full of things I'd made, Jamie hired me and I worked for him for the better part of the next four years. Jamie was the opposite of the guys in Manhattan—he made his entire shop available to me for anything I wanted to learn. He helped me grow as a maker more than any single person I encountered before or since. He also helped me realize that I had a passion for model making and practical effects.

The effects community is fairly insular. After a few years steeped in it, you get to know pretty much everyone in your field. In San Francisco, that meant you crossed paths with the artists at Industrial Light & Magic. By 1997, I'd become friends with a half-dozen artisans in their model shop, and shortly thereafter they told me ILM was hiring for the model shop and that they'd put in a good word for me to help me get a leg up. I called their boss, Rick Anderson, every week for three months until he agreed to bring me in. I spent the next few years living my childhood dream working for ILM.

Around 2002, when Jamie and I were simply professional colleagues who stayed in contact, he got in touch about this new show called *MythBusters* that Discovery Channel had ordered from an Australian production company. They wanted Jamie for the host role, and they'd asked him to make a demo reel, but he didn't think he could host a show himself—*he needed help*—so he thought back through the catalog of people he'd worked with who were both good at making stuff in a variety of forms and also potentially good on camera. (I believe the word he used was "hammy.") I was the first person he thought of. He needed help, I had help to give. We ended up having a perfect funny man/straight man dynamic, and together we could build most anything, including a show that would end up running fourteen seasons, spawn a reboot and a kids' version, and open the doors to collaboration on a digital portal for makers, culminating in building a science-based stage show that toured over two hundred theaters on three continents.

HOW DO *WE* GET THIS DONE?

I thought I'd licked my "asking for help" problem by the time Jamie and I joined forces for *MythBusters*, but really the problem had just shape-shifted like an Icelandic berserker, intent on fighting to keep me firmly planted in my own way—this time as a boss, running my own shop both on the set of the show and then down in the Mission District of San Francisco at Tested HQ. Part of being the boss of a shop is delegating basic process tasks to your team members so you can focus on the more important, high-level stuff that keeps a shop running: meeting with clients, coming up with big ideas, paying the bills, that sort of thing.

Sounds easy enough, right? It is certainly straightforward. Unless, like me, you work really fast and love the sense of accomplishment that comes with completing tasks yourself and checking them off your list. Then basic delegation isn't so simple. If I can do something in an hour, that a less experienced maker in the shop might take two or three hours to complete, the question I ask myself isn't, "How can I help them do it faster?" It's "Why not just jump in and *do it myself*?" Then I invariably do, and the first question never gets answered.

It is this tendency that has made me a poor delegator. It's also made me less productive in the past. My big problem is that I want to do too many things, and if I load them all up on my plate because I know, individually, I can complete each task more quickly than anyone else, the end result is that nothing gets done and, thanks to my chronic impatience, I haven't helped my younger collaborators learn new skills to make them better. The shop then naturally grinds to a halt. Tasks get completed, sure, but not whole builds. Nothing goes out the door, and unfinished projects start to crowd out new opportunities. This means I HAVE to ask for help from my team. I have to delegate. It's a crappy feeling to fight with this aspect of creation. This makes it sound like I don't trust my collaborators. But really, it has nothing to do with them. It has everything to do with me.

I am not alone in this struggle. Besides being a fellow obsessive, Jen Schachter is a master maker, and a deeply systematic thinker whose entire professional mission is to equip the next generation of creators with the tools to do what she does, and yet in her own work delegation can be a Herculean task. "One of the things I struggle with is sometimes it's easier to do something yourself than to train somebody else to think the way that you think about something," she said. "I don't want to take the time out of what I'm doing to have to bring somebody else into the fold."

I know the feeling. It's so hard to make other people understand not just your methods, but also how you arrived at the conclusion that these are the best ways to complete particular tasks. I've been doing this for four decades now. There were so many twists and turns in the path I took to get where I am as the head of a shop, I couldn't possibly transmit all that backstory, all that rationale, to anyone no matter how long you gave me. I can tell them what to do. I can tell them how to do it. But I can't tell them how I know that's the best way to do it.

Jen understands this problem, too. "I am so particular about the way that I do things," she told me, commiserating on the phone over this issue, "that I almost feel like, 'Well, no one else will do it the way that I'm going to do it.'" Her voice trailed off. She didn't finish the sentence. Almost like she knew that I knew what she was going to say, and we *both* knew it wasn't the right attitude for running a shop. She was going to say, *So I might as well take care of it myself.*

The irony is, when we can take our "boss" hats off and think simply as makers, we both love the art of collaboration. In fact, Jen isn't just a friend, we are also frequent collaborators. One of President Obama's last celebrations at the White House was called South by South Lawn, a reimagining of Austin, Texas's famous SXSW conference, on the south lawn of the White House. I envisioned a giant lit-up sign, and Jen designed it. Together with fifty kids from the Digital Harbor Foundation in Baltimore, we spent fourteen hours putting together Jen's incredible design, and then the next morning I got to do something that I never thought would be on my list of coolest life moments: with Jen, I drove a U-Haul van ONTO the White House lawn. The honeymoon did not last long. We were immediately, yet politely, told by the Secret Service that we had eight minutes to unload, assemble, and get the hell out, which we did, with a minute to spare. It was exhilarating.

"There's a level of transcendence that you can get to when you're working by yourself on a project and everything's sort of vibrating and you're just in the zone. But, it's a whole other level when you're doing that with other people," Jen explained as we reminisced about our White House collaboration. "It's super, super satisfying. You're all on this other plane of existence together and you can get to a point where you communicate without words. You just look at the other person and they anticipate what you're trying to say and hand you the thing you were thinking about."

This was one of the things I loved best about working with Jamie Hyneman. We are very different people out in the world, but within the four walls of the *MythBusters* workshop Jamie and I could build almost anything together, communicating solely with pointing and sketches and simple pronouns.

For Jen, the joy of collaboration extends beyond this transcendent synergy to something even more elemental: "There's something really gratifying about building something that you couldn't possibly have made on your own because of the sheer scope of the work, but more so because you don't have all the knowledge to do it. My biggest satisfaction comes out of projects that have required lots of people's expertise and hands and labor to make happen. Then we can look at the finished product and say that we did this together, that this is not just something that I made."

THE ACKNOWLEDGMENT HIERARCHY

As bad as I have been at times about asking for help, the aspect of collaboration that came to me latest in the game is perhaps the most necessary to perfect: acknowledgment and appreciation, that is, acknowledging the help that has been given to you—by people you work with or work for—whether you asked for it or not. This is especially true when you are the boss.

We filmed *MythBusters* for fourteen years largely with the same crew, many of them matriculating up from being runners and interns to being key producers, camera people, and creatives on the production. We operated as much like a family as we did like a television show and a business, which is to say that we all saw each other perform at our absolute best, as well as at our worst. Like anybody, I had good days, and occasional bad days. My bad days, though, were not localized to my own experience. As one of the bosses on the show, they rippled out into the production and tended to affect the mood of the entire set.

Invariably, I could trace the source of my bad days back to how I managed (or didn't manage) the work going on around me. The problem was never specifically about delegation or collaboration, which is why it took me so much longer to get the picture. It was more about communication. Crew members would make the same mistake multiple times. Things wouldn't get done to Jamie's or my specs. Production would bog down. I wouldn't know why and then I'd get frustrated, but because I am so nonconfrontational, it would take forever to get to the heart of issues that should have been molehills but that I had allowed to grow into mountains.

What am I doing wrong? Why won't this person ever learn?! It was a regular refrain in my inner dialogue between takes. Then one day I realized that all the most valuable stuff I ever learned as a maker came in the form of critical feedback from employers or clients—and I was providing almost none of that to my team. I'm naturally allergic to telling people things they don't want to hear, but by looking at my own past, I saw how necessary it is to give proper, contextual feedback to the people you work with—to acknowledge their work, to appreciate their effort, and to correct their mistakes.

Personally, I believe there are levels to this kind of feedback. It is a hierarchy of acknowledgment that moves from positive to negative, and gets harder and more important, the deeper you get. At the top level is simple gratitude: saying "Good job! Well done!" when the situation warrants it; saying "Thank you" when someone offers a hand. It's not something that requires public pronouncement, it's just basic common courtesy.

Below that is encouragement: letting the people who have helped you know why their work was good. "Hey Tory Fink!" I might say to *MythBusters* lead builder, "I can tell you had fun building that prop, and it not only shows, it made the whole episode better, thanks for killing that and showing me a new side of what you can do!" That might sound corny to a more cynical ear, but when you're working twelve- to sixteen-hour production days for weeks at a stretch, people need to know that their effort has been recognized and appreciated by the people writing the checks.

Then there's motivation: giving someone the context for why they in particular are perfect for the role they're performing; explaining how they contribute invaluably to the whole picture; and reminding them that you couldn't do it without them. Beyond encouragement, which is recognition that all this hard work is not in vain, motivation is about getting more out of someone when it matters most.

There is a point however when feedback tilts from positive to negative, beginning with basic constructive criticism. This is the part of the curve I have the most trouble with. Constructive criticism usually seems simple—it's just giving guidance when something isn't working or a project needs a small change of direction that a specific person is responsible for implementing. But it's often tricky because to you it's rarely that big of a deal, but to the recipient it can be this short little burst of negativity. Nobody likes hearing that they didn't do a complete job, but being able to give honest feedback is critical to working with others. I don't like telling people things they don't want to hear, but I remind myself that by providing negative feedback, I'm investing in the people I'm talking to. If I didn't think they were any good, I'd not bother telling them they did something wrong, I'd just not hire them again. Keeping that in mind has helped me give better critique.

Below that is a larger course correction: a project is going way off course and we really need to marshall our forces to get it back on track. Usually there's a point person who is principally responsible for the part of the project that's gone off the rails, but a full course correction is never just about one person. Just as it only takes one person to derail a train, but an army to get it back on the tracks, when you've reached this point with a project your feedback is going out to the entire team.

Lastly and most importantly of all, is confronting someone about a personality trait that is inhibiting things from moving forward smoothly for everyone else. This is truly difficult for me. As I've said, I don't like telling people things they don't want to hear, but I love working with my team and I don't want to let anything or anyone ruin that collaborative environment for them. See! I am so nonconfrontational I'm basically talking about having to fire someone and I don't even want to type the word "fire" as a way to describe it.

Each of these levels on the acknowledgment hierarchy is vital to get right, whether you are working with friends, family, colleagues, vendors, bosses, or clients, because this is one area where you cannot ask for help. When you're the boss, when the buck stops with you, then you need to be able to deliver the right feedback at the right time. It might be new to you, as it was to me on the set of *MythBusters* those first few seasons, but the terror that comes with newness can't be your excuse for falling back into old habits or retreating to the comfort of the way you've always done things.

That is the cruel joke of making, or any creative discipline for that matter: no matter how much you progress in your career, the duality of thrill and terror that exists with all new things will never leave you. In fact, the better you get and the more experience you acquire, the better you are at understanding objectively where your work and your wisdom fall short. Self-doubt never leaves the attentive craftsperson, so you best make friends with it. And if you don't know how, here's some advice: have the patience and humility to ask someone who's been there before.

DEADLINES

We don't do well with time. We struggle to manage it, to take advantage of it, even to conceptualize it. When we have to get something important done, we often feel like we either have absolutely no time or all the time in the world. Either end of the spectrum can hamstring us. We feel crippled by too much or too little leeway, and then, nothing gets done.

We have several different names for this phenomenon, depending on how it manifests: procrastination, perfectionism, analysis paralysis, Hick's Law, the paradox of choice. Whatever you want to call it, this tendency is the bane of the maker's existence. More specifically, it is the bane of *my* existence as a maker if I don't do something to mitigate it.

That something is almost always to make deadlines. Everything in the previous chapter was about ways to be more efficient and effective, how to ask for help and offer it to others. Deadlines are about helping *yourself*. I LOVE DEADLINES! They are the chain saw that prunes decision trees. They create limits, refine intention, and focus effort. They are perhaps the greatest productivity tool we have, and you don't need a Time Life series of books to learn how to make them.

Deadlines are also excuse-busters. They breach the walls we put up between ourselves and new, unfamiliar things that deep down we really want to try. As makers, these walls are the excuses we create for not having created, for not starting or doing or working: *I don't know what to make; I don't know how to make it; What if I mess up?; I don't have everything I need.* Self-imposed limitations on your time can be the blast charge that crumbles the barrier between you and creativity. They aren't the most fun things in the world, and they take some getting used to, but when you're sitting at your workbench or your desk and you're looking at a finished project, you will almost certainly have deadlines to thank at least in part. You just have to be willing to embrace them.

A COUPLE MORE WEEKS AND THIS COULD BE NICE

When I first got to Industrial Light & Magic to work on *Star Wars*, somewhere in the back of my mind I secretly hoped that walking through those doors every day would be like the model maker's version of Willy Wonka's Chocolate Factory. The walls would be a dripping kaleidoscope of paint in every hue on the Pantone color chart. Shake a tree and all types and sizes of bolt, nut, screw, and fastener imaginable would fall at your feet. Behind every door new creative wonders would emerge. And to an extent, that was the case. At ILM, there was never any shortage of tools, materials, or expertise. Only time. No matter how much of it you had on a job, there was never, ever enough.

One of the first projects to which I was assigned after *Episode I* wrapped was a little Clint Eastwood movie called *Space Cowboys* about a group of old astronauts who get plucked out of retirement to go up into space to rescue a wayward satellite. I say "little"

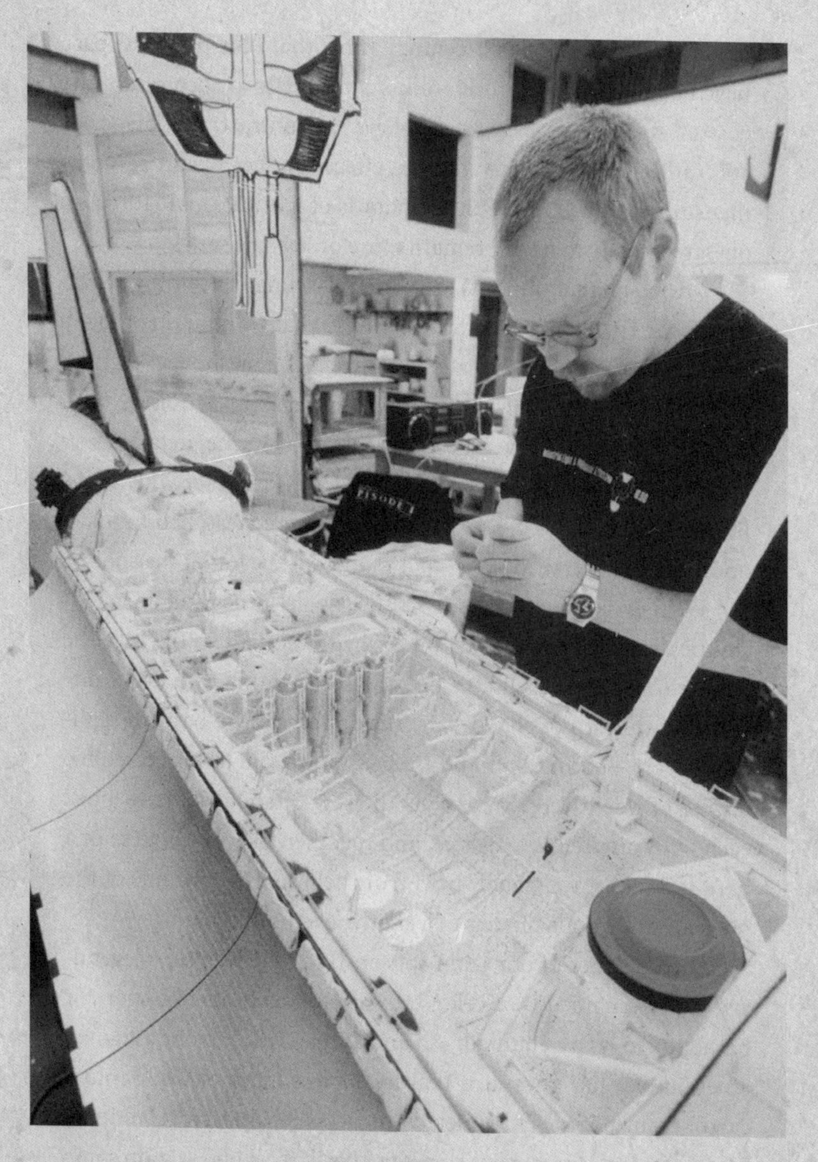

My main job on Space Cowboys *was the construction of everything in the payload bay. What a dream that job was, laboring over every nut and bolt inside the doors of the shuttle.*

because most twenty-first-century films that take place in space have budgets well over $100 million, primarily to account for the cost of visual effects. *Space Cowboys*, by contrast, was made for $65 million. This meant that the effects budget would be commensurately lower, while the standards of fabrication quality and on-screen realism would remain more or less the same.

My job on that film was to build everything inside the payload bay of the fictional NASA Space Shuttle that Clint and company would take up to save the satellite. Were this a big-budget picture, with all of the shots of the shuttle that the script required, we'd normally have built several different sizes of shuttle, detailed sections for close-ups, and less detailed bigger sections for wide shots. But with *Space Cowboys*' smaller budget, we instead had to make one incredibly detailed seven-foot model suffice for all the shots.

With orthographic layouts of the tile patterns used on actual shuttles supplied by NASA,* we scribed every single tile into the outer surface, and then attached the serial number to every one with the tiniest possible decals. Inside the shuttle bay, we replicated every nut and bolt and hinge and handle to an incredible degree of fidelity. We were a small crew of about seven people, each with different specialties, and together, over the course of a hard ten weeks, we made what I truly believe might be one of the most accurate scale models of a shuttle ever built.

One member of our team was an ILM veteran and a legendary sculptor named Ira Keeler. Ira was famous inside the shop for his ability to take a chunk of ash and a small finger plane and over a period of a few days turn that featureless hunk of wood into a car, or a storm-trooper helmet, or an airplane wing. He'd laminate pieces of wood together, make major cuts with a band saw, do some

* Orthographic projection is a way to represent three dimensions in two-dimensional space.

rough shaping with chisels and rasps, then with the small plane not much longer than his index finger, he'd carefully shave away tiny chips of wood until he'd removed every single bit that didn't belong—until he'd freed the angel from the marble, to borrow a phrase from Michelangelo, the most famous sculptor of them all.

It was awe-inspiring to watch him do this kind of thing up close while we worked together, but what will always stick in my mind is one of his favorite catchphrases. He'd be sanding, shaping something like a beautiful curve between the body of the shuttle and its wing, step back, scrutinize it for a moment, then casually say, "You know with a couple more weeks this could be a nice model . . ."

Ira had been saying that for decades. It was like hearing a Zen koan from the *Caddyshack* version of the Dalai Lama. It always made us chuckle. It's such a relatable sentiment. EVERYONE continuously wants more time, and it's always the deadline that is the problem, not the budget or the size of the workforce. We had ten weeks, which seems like a long time, but for one model that you can't screw up, with this level of granular detail, and only seven pairs of hands on deck, ten weeks was a tough sprint. Yet no one ever said, if only we had a couple more hands or a couple million more dollars. No, they said, if only we had a couple more weeks, just imagine what we could have done with this thing!

This is exactly the trap you don't want to fall into when it comes to deadlines: you don't want to cast them as the villain. What you want to do is embrace them, because at a certain point more time does not equal better output. In fact, I think not having enough time is critical to making and, most importantly, finishing things. You know what we wouldn't have done if we had two more weeks on the *Space Cowboys* shuttle? We wouldn't have been finished with it. We would have wanted *two more weeks* and then our next big project would have been pushed out a full month. Thankfully, this is something I already understood from my time working on commercials with Jamie, so I was comfortable embracing the metaphorical hustle across Wonka's factory floor, instead of resisting it.

WHAT'S THE POINT OF ALL THIS?

While I worked for Jamie at Colossal Pictures we did about one commercial campaign every few days, without accounting for vacation time or sick days or utter exhaustion. The lack of time we had on every project was nothing short of stunning. The pace was grinding, grueling, and completely exhilarating. It was also incredibly formative, because the deadlines under which we operated, in all their various forms, taught me how to figure out what was most important.

Early on in my tenure, we worked on an ad campaign for Toys "R" Us. They were shooting twelve separate ads over a three-day stretch (which is insane, if you were wondering), each one featuring its own unique special effects rig. My favorite was a pen that had to look like it was flying around in the hands of a ghost. During the shoot, I spent an hour and a half contorting myself just outside the frame, puppeteering this tiny pen from nearly invisible wire to make it look like it was writing. It was excruciatingly difficult to get just right and the director, Carl Willat, could tell. So between shots, as the crew reset a shot, he came over to me and said reassuringly: "Remember, Adam, you have to get this right because each and every take costs *thousands* of dollars." Thanks, Carl. He was joking, but not really. On any filming set, time really is money. And with limitations on both, there always comes a huge amount of pressure to get it right on the first go.

Of course, then things on the Toys "R" Us set only got worse, this time for Jamie. One of the rigs on a different ad involved a sensitive spring-loaded magic-trick-like gag that he had been laboring over for the better part of a week. When it came time to set it up for filming, a key component exploded into three separate pieces. It was immediately obvious to Jamie and me that the rig was DOA. There was no way, under any circumstances, that this thing was going to work for the shoot in the time allotted.

This posed a serious problem. For one, there was now a crew of forty people standing around, getting paid to do nothing, while they waited for Jamie to make this thing work so they could finish their day. But more than that, Jamie had contracted with Toys "R" Us and the production company to produce *these* special effects for the ad campaign and now one of them just wasn't going to happen, which meant technically, he could have been in breach of contract.

Jamie's response to this incredible pressure was both surprising and inspiring. He didn't show any emotion, neither perturbation nor anger, not even nonchalance. He just calmly looked at the producer and said: "To get this done by the end of the day I figure we have three options . . ." Then he carefully laid out three brand-new solutions, complete with the pros and cons for each as they related to the original storyboards. In retrospect, this was Jamie's only choice, but in the moment, the presence of mind he displayed in presenting these options to his collaborators showed a profound understanding of the nature of deadlines and a full embrace of their role in the creative process.

Just think about the position Jamie was in: his rig was not going to work; he was under contract to deliver these special effects; and there were only three days to shoot twelve of these things. There was no time for apologies or excuses or emotions. This thing *had* to get done. He just needed to figure out what his options were and let the producers pick one, before moving forward, which is exactly what they did. The producers picked one, Jamie set me to work on getting it done while he rigged the next commercial in the queue. We eventually got the gag to work, and all was right in the world by the end of the day.

My friend Dara Dotz knows all about this particular kind of pressure. Dara is the cofounder of a humanitarian organization called FieldReady.org. She and her team of designers, engineers, and makers deploy to disaster areas and sites of humanitarian crisis armed with various rapid prototyping technologies to help solve pressing infrastructure problems like broken plumbing and irrigation systems, makeshift housing, downed electrical grids, and impassable roads. They design the solutions, build them, and then share the knowledge with the locals so they can maintain the systems once Dara and her team have gone. They are like guardian angels with hardhats and tool belts and CAD programs. Every single minute of their work is done against the natural deadline of life or death.

"That's my regular life," she told me over the phone one day after she'd just returned from somewhere overseas. "In a disaster, it's like 'Oh shit, we gotta get something right now.' Then after that stuff we do right then and there, it's like, 'Okay now we just need something to get them to the next point.' "

Moments like these are when many people freeze. This has been the hardest of many hard lessons Dara has learned in her time growing FieldReady into the organization it is today. "The biggest thing is figuring out which people to invest in," she said. "Some people can do it and some people can't and that's just part of it." The reason some people can't do the kind of work Field-

Ready does is the same reason some makers get paralyzed by looming deadlines: they internalize the consequences. Failure to meet the demands of the time pressure is their failure, which means that they are failures, themselves. The existential weight of that becomes too much to bear. There is a fair amount of ego in that perspective. To think that the trials and tribulations of a complex build, whether in a model shop or on a commercial film set or in a disaster zone, reflect wholly on you as a maker or as a person is a bit much, no?

Like Dara, Jamie was immune to that kind of negative thinking because he understood why we were all there and what the point of all the effort was. The forty-person crew wasn't there for our special effects rigs. They were there for the commercials. And the commercial project was there for Toys "R" Us. As model makers and special effects guys, of course we would have preferred to take the time we needed to make the original rig work perfectly, just the way we designed it. But we were not the star of this show, we were bit players on this particular stage whose job it was to help move the plot forward, so to speak. And it was the grinding pressure of a hard deadline that helped Jamie clarify his decision making in that regard.

The clarifying power of a deadline is something that every maker should embrace in their own work. We should be asking ourselves repeatedly, "What is the essence of this project?" as we move down the path toward completion. And as the delivery deadline nears, we should ask that question more frequently, because it helps us remember why we're there, and what the point of the whole project is. This is especially true if you're working on something for yourself, because then it can become anything. And if it can become anything, then what is it really? A deadline shouldn't feel like a vise slowly crushing your head, it should feel like a sieve through which only the essential elements get pushed by the pressure of time, leaving the unnecessary bits behind.

In a sense, this is the maker's version of the End of the World question: If the world was going to end tomorrow what would you do today? For makers like Dara Dotz, those questions are actually very real and have immediate consequences. For the rest of us mere maker mortals, the answer to each of those questions tells us something about what matters to us, about the nature of our projects themselves, about what the point of all our work has been.

JUST MAKE IT WORK

Working on commercials week after week, day after day, every single hour a new set of problems to solve, was my actual college education. I had learned a huge amount already in theater prop work, but this was graduate-level studies. Aside from helping me clarify the larger goals of my work, commercial work taught me how to constantly play through worst-case scenarios, because when a crew that costs thousands of dollars a minute is waiting for YOU to finish that thing YOU'VE promised them would work, you don't have time for indecision or equivocation. You only have time to make it work and then move on to the next problem.

In 1997, I had the opportunity to make a Coca-Cola television commercial with my friend, the amazing art director Lucy Blackwell. The conceit of the commercial was that someone buys a Coke from a futuristic vending machine, and we're taken inside the machine to see how the Coke bottle is selected, opened, poured into a glass, and then delivered to its thirsty purchaser via a ridiculously elaborate Rube Goldberg machine.

There would be marbles that bounced down stairs made of

hand scoops; eggs released from mechanical baskets to force counterweighted cans to roll uphill; shuttlecocks dropped into bowls that flipped over tuning forks, which then released three ice cubes from a frozen tube that would drop one by one onto a latex drumhead, which would then bounce precisely into a glass twenty inches away—all three in the same shot—while the last marble from the very first shot came rolling down a staircase made of bottle caps to trigger the opening and tipping of the bottle, which was then caught with a rubber glove that was instantaneously inflated to properly pour the Coke into the waiting glass of ice.

All told, this Rube Goldberg machine of ours would include more than a dozen mechanical set pieces, each requiring design, engineering, problem solving, and then execution on set, over two long days of filming. The client who hired us only had two conditions: 1) all the gags had to be practical, which meant everything we see in the commercial actually had to work in camera; and 2) we had seven weeks to get it done.

At the beginning of the job, seven weeks seemed like an impossibly luxurious amount of time, which allowed us to indulge in our choices. What kind of marbles should we use? What size should they be? How many? Should they be Coca-Cola red or a mix of colors? It felt like we could be that particular with all our choices. But as the shoot days approached, the sense of optionality melted away, and it began to feel like we had barely enough time to finish the fabrication piece of the build alone.

On set, the sense of the clock running out on us continued. It culminated with a tense hour near the end of the last day when the ice cube–bouncing setup repeatedly failed. The elements of this sequence were fairly straightforward: bounce three ice cubes, in succession, off a stretched latex drumhead, so that they land

perfectly one on top of the other in a classic Coke glass, all in the same shot. Simple enough, just not easy. We couldn't use real ice because the intense commercial lighting would melt the ice cubes in a hot second. We had beautiful *fake* ice, but it, too, wouldn't work because each cube was slightly different, making their bounce trajectories impossible to unify. Fortunately, in Jamie's shop I found a box labeled "fake food" with a handful of perfectly identical clear acrylic "ice cubes" inside, which, if I lined them up in the exact same orientation, would land where I wanted.

It was the drumhead that ended up presenting a trickier problem: the hot stage lights were heating up the latex and changing how bouncy it was. The "ice cubes" would hit the glass in rehearsal, but when the lights came on during filming, they'd miss. I was in a jam. The perfectionist and the completist in me wanted to steal back some of that seven weeks we'd used to find the most elegant, most aesthetically pleasing, most creative solution possible. Having already learned from my time with Jamie, though, I knew that the most important thing was getting the commercial in the can. To do that, I had to strip away all the extravagant choices and elegant solutions, all the wouldas, couldas, and shouldas that come with second-guessing how you got yourself into this position, and figure out the simplest way to make the thing work.

I remember several members of the film crew staring at me in disbelief, shaking their heads as I stood there racking my brain. They had no faith that this gag was possible, and they had no problem letting me know. I wanted desperately to find a mechanical solution to this problem, but pretty quickly I realized that with the distance of the bounce constantly changing on the deforming drumhead, I was going to have to take matters into my own hands, quite literally. On every single take, I manually and minutely adjusted the position of the drumhead to dial in the distance and ensure that the ice cubes hit their target and landed in the glass when the lights were on and the cameras were rolling. In the end, I was able to facilitate five good takes. For a gimmick built around practical effects, it doesn't get more practical than that.*

That is what deadlines can do for your creative thinking. They help you cut through the clutter. They trim all the pretty branches on the decision tree that sidetrack you or can't hold your weight as you try to make it to the top. Just think, with any project you probably have twenty-five or thirty discrete things you want to achieve by the end of it, but it's never totally clear at the start how many of those are in service of the project and how many are simply personal creative preferences. With no limitations on your resources, each one of those goals is liable to get equal weight and attention from you, potentially extending the life span of the project for years. With the help of a deadline, however, those goals start to change size and shape in proportion to one another and to the project as a whole. As the deadline approaches, the ones that don't impact function shrink and narrow and become ancillary, and fall away. Then, before you know it, you have a finished product that embodies ten or twelve of your original goals, each of them perfectly suited to the role they fill and essential to the project.

---

* https://vimeo.com/35240952 I'm amazed to this day that they let this commercial on television, it's just so weird and awesome. It still stands out in my mind as my favorite commercial I ever worked on.

On the set of the Coca-Cola commercial, my desire for real ice cubes and a beautiful, smooth surface to bounce them off were really just personal preferences. What was important was that I needed to make the ice cube gag work, and it was the encroaching deadline that bounded the universe of possibilities at my disposal so that everything else fell away and the proper solution made itself known. As a maker, you need to think about employing deadlines on your projects this way, whether you're making something for yourself or a client, and particularly if you are a compulsive perfectionist and prone to going deep down dark rabbit holes. Deadlines can help you focus your attention on the elements that are most important to your project's existence and its purpose, so you don't become one of those tortured artist clichés who let perfect become the enemy of good.

PERFECT IS THE ENEMY OF DONE

Deadlines are often the only way I can get a project done, especially if I'm following a particular thrill that leads me to make something just for myself. When I was working on *MythBusters*, our schedule was intense. Most television shows film over a period of a few months, and then they have an off-season. We didn't. In order to meet the delivery demands for Discovery Channel, we filmed forty-two weeks out of the year, in three-month blocks, with two weeks off in between. As a result, even in a good month I almost never got to spend more than five to ten hours in my own shop, working on my own projects. In bad months (which outnumbered the good), I had so little time that some of the projects I was working on dragged on for years while I pecked away at them. One project in particular, a space suit from *Alien*, lagged for almost a decade and a half before I finished it, and it was only setting a deadline in order to have it ready for Comic-Con that finally got it across the finish line.

I've always loved Ridley Scott's *Alien* for, among many other stellar qualities, being some of the most effective world building in the history of science fiction. Ridley and his Academy Award–winning production designer, John Mollo, built a world that was so tangible, so viscerally real, where every piece felt like part of a cohesive whole, that I could imagine myself stepping on to the commercial refinery vessel *Nostromo* and knowing exactly where to go and what to do when they receive the distress signal that is the inciting incident of the story and head down to the surface of an unknown planet to investigate.

Unlike the silver-suited utopia promised by the sci-fi films of the mid-twentieth century, *Alien* is working-class science fiction. As such, the space suits that the *Nostromo* crew wore to track the source of the distress signal are my favorites in the history of film. Designed by legendary artist Moebius (aka Jean Giraud) during his brief stint in the *Alien* art department, they're a master class in my favorite kind of storytelling. Worn and patinaed by use, they're a cross between early deep-sea diving suits and samurai armor. A patchwork of retrofitted pieces and discordant details, they communicate instantly that this is gritty, roughneck sci-fi. There's no romanticism to be found here, just blood, sweat, and space dirt. So, of course, I have always wanted one.

In 2002, I began working on my own replica of the suit worn by John Hurt's character, Kane. For months I gathered pictures, scanned old magazines, and collected information, anything I could find. I was lucky enough to spend a few hours with one of the real costumes (the one Veronica Cartwright wore) and make measurements and drawings that enabled me to do much of the problem solving for the soft parts of the costume. This was critical intel to have before embarking on fabrication of the hard parts, which were, true to their name, very hard. All of this took place over a period of about three years, and I wasn't even close to finishing.

From there the project lagged on and on. There were so many small parts, so many details to wrangle. I had two large binders filled with information, pictures, plans, close-ups, blueprints, my own drawings, and lists upon lists. I had (and still have) a couple gigabytes of reference material on an external hard drive somewhere. When *MythBusters* wasn't taking up 80 percent of my time, I would dive into the project when the mood struck me, but never for long enough to build the momentum I needed to finish.

Then, starting in 2013, our workload on *MythBusters* got cut in half. We went from doing two dozen episodes a season, as the series began to wind down, to anywhere from ten to fifteen. This gave me more time in my shop and it freed up more creative space in my brain. It also helped me realize that I had to do something differently if I was going to finish this suit once and for all. It had been long enough. I needed to give myself a deadline. I found one in the form of San Diego Comic-Con in July 2014.

I've been going to comic conventions for well over a decade—sometimes on a panel, sometimes to satisfy my love for cosplay, oftentimes for both. Usually, I'll build and wear an elaborate costume on the floor of the 'Con and have a scavenger hunt for fans to find me. The first time was in 2009 and I went dressed as Hellboy. The next year, I wore a *Star Wars* storm-trooper outfit that we'd used in a *MythBusters* episode. In 2011, I went as No-Face from *Spirited Away.* In 2012, I was the Ringwraith from *Lord of the Rings.* In 2013, I was Jack Sparrow from *Pirates of the Caribbean* AND Admiral Ackbar from *Return of the Jedi.* Leading up

I have never been so buff in my life.

That's me wearing Totoro on Halloween 2016 in my neighborhood of the Mission, San Francisco. My wife, Julia Ward, took this photo. It's one of my favorites.

to the planning for 2014, I knew that I wanted to complete Kane's *Alien* space suit and wear it on the floor at San Diego Comic-Con later that summer.

I'm still not sure how I got it done, but I know that if I hadn't set the deadline my completist side would have been content to continue obsessing over every little thing, and never getting anywhere. I would have agonized over which lights were best, which cooling fan was most authentic. I would have fixated over things that nobody but me would EVER notice. With the July deadline, the luxury of that kind of indecision was eliminated.

There is a key to setting a deadline for yourself in a situation like this. It cannot be totally arbitrary. The deadline has to be relevant to you or to the project, or both. Comic-Con was the ideal combination: I was booked to go and the costume was perfect for the occasion. I could have set a December 31 deadline for the *Alien* space suit, but the end of the year is a delineation that didn't connect to me or the project, so I could have easily pushed it off. I could have picked Christmas, and that might have worked if I was going to give the suit to someone as a gift, but again this project was just for me.

When your project is starting to lag, figure out a date that is as important to you as the project, then work backwards from there. Trust me, you'll get it done.

IT'S NOT THE END OF THE WORLD

Perhaps the most important lesson I learned about the intersection of deadlines and creativity came from a big-budget Super Bowl commercial I worked on for Corning with a brilliant model maker named Mitch Romanauski. Mitch worked on *Top Gun*, *Pirates of the Caribbean*, *Star Wars: Episode II*, and *Jurassic Park III*. He supervised the massive team of artisans who built literally everything for Tim Burton's classic stop-motion film, *The Nightmare Before Christmas*. Mitch was a genius. He partnered with Jamie after he left *James and the Giant Peach* and became my first professional model-making mentor. Jamie was my boss, but Mitch really taught me the ropes.

On this Super Bowl commercial, we worked for a director whose preferred mode of communication was yelling and blaming. We were also given an extraordinarily short amount of time to complete our work, even by low-budget commercial standards, so there were a number of very long, tense days trying to check things off our to-do lists.

The day before the shoot, Mitch and I ended up pulling an all-nighter to get all the bits and pieces done so they could start rolling by eight the next morning. In the moment it felt like an overwhelming amount of things to complete: stretching latex walls with precise printing on them; miniature housing insulation that rolled itself; a building that assembled itself board by board while the camera watched. And that's just what I can remember, besides the fact that we were absolutely killing ourselves.

I smoked back then. So did Mitch. We weren't allowed to smoke in the shop during the day, but once the sun went down and everyone else went home we chain-smoked our way to wakefulness to get the job done.* Mitch and I managed to avoid any significant screwups as we marched toward the morning, but not without some gallows humor to buoy us. With the pressure bearing down, sometime between 3 and 4 a.m., Mitch turned to me and said, "You know it's around this time of night that I get real philosophical about this kind of crap. I mean, at the end of the day it's

* Sorry Jamie, I know we were violating OSHA rules, but war is hell.

just another damned commercial, and we'll either get our part of the job done or we won't. It's not like the world's going to end . . ."

Mitch was absolutely right.

I would end up making a week's pay in two days for the overtime I pulled on that job. By the time it was all over, I had worked through the night and punished my lungs. I'd been yelled at by the director. I got blamed by the animator when he screwed up his part of the job and I had to take over and, for twenty-five straight hours, physically animate the house building itself. And yet, I remember working on that commercial with nothing but fondness and affection, because we got everything done in spite of all the nonsense and external pressure. We completed our work on time, I believe precisely because it wasn't the end of the world if we hadn't. That understanding took the anxiety and existential dread out of the deadline, which had motivated us to work as hard and as effectively as we did.

Knowing that missing a deadline wouldn't be the end of me as a maker freed me to channel all the stress tied up with a ticking clock into a focused perspective on productivity. It gave me the skills to solve complex problems while keeping my eye on the ultimate goal of the project, whatever it might be that day. This is where list making becomes an invaluable tool, by the way. List making is about momentum, especially when I'm on the clock but don't want to feel like I am. I need to know how what I'm doing will fit into the time I have. If I spend too long on one part without realizing that I have twenty steps *after* that, and only an hour to go, my momentum on one thing will kill momentum on the overall project, and shoot me way past my deadline. A list keeps my decision tree bearing fruit, and keeps me out of time-sucking dead ends. Make no mistake, these are all tools I employed and lessons I applied originally in order to get stuff done in my commercial work and then on the set of *MythBusters*; but it was once I transferred them to my personal work, as a replicator and cosplayer, that the benefits really started to accrue. The result was nothing short of a radical, miraculous increase in productivity . . . and joy.

DRAWING

Do you have an idea in your head? One you're proud of and excited to build? Okay, I want you to try an experiment. Call up a friend, maybe even a collaborator, someone who has at least some creative ability in their bones, and describe your idea to them. Tell them what it does, how it works, what it's made of. Describe its shape and size and color. Explain how you're going to make it: where you're planning on drilling the large bore holes, what end you plan on flattening, which end you want to chamfer, what you're going to use for final assembly. Give your friend all the detail you can.

Now, ask your friend to draw with paper and pencil what you just described. Give them all the time they need, answer any question they might have. In my experience, it won't matter, because what your friend eventually produces will look so radically different than the image you tried to describe you will wonder if they were listening at all. The differences will shock you. You will wonder if you were speaking a foreign language. The divergence between the image in your head and the image produced by your friend may even be enough to make you question your whole idea (please don't!).

But if *you* can draw it? If you can successfully demonstrate your idea in a way where both of you have the same understanding of it, so that either of you could build it or use it? That is a worthy goal. In my experience, the ability to take an idea from your own mind and transfer it to the mind of another person is intoxicating. It is a kind of creative empowerment that makes all your other crazy ideas feel maybe not so crazy. And the fact that you only need a pencil and a piece of paper to make it happen, that is most empowering of all.

Despite the fact that I haven't, until quite recently, felt particularly skilled at it, I've been drawing most all my life. I draw every single day, for a multitude of reasons. I use drawing to flesh out and perfect my ideas. I use it to communicate with other builders and colleagues. I use it to create momentum. I use it to capture knowledge developed over the course of a project. And, of course, I use it for ideation.

From a planning perspective—whether it's for current or future projects—I look at drawing as a translation tool from my brain to the physical world, where I have frequently found words wanting in the explanation of complex objects and operations, which, of course, is the entire purpose of every plan ever made. What is a plan if it isn't helping you understand what you're building and how you're supposed to build it?

Today, the maker space is not lacking in planning tools. There are software and mobile apps and various mechanical apparatus, and they all work the way they're designed, but none of them seem to do what a simple pencil and piece of paper can. Because unlike those other methods, drawing out your idea shares the physical, tactile character of the building and making it is meant to precede and facilitate. Drawing is your brain transferring your idea, your knowledge, your intentions, from the electrical storm cloud at its center, through synapses and nerve endings, through the pencil in your hand, through your fingers, until it is captured in the permanence of the page, in physical space. It is, I have come to appreciate, a fundamental act of creation.

DRAWING AS RENDERING

I grew up with a beautiful 1905 Brunswick Balke Collender pool table in my house. I played all the time when I was living at home, but not seriously, mostly just mucking around. I knew a ton of games, I could occasionally apply some useful English and backspin, I could trash-talk with the best of them, and I could hit the occasional bank shot, but I never really got good. It was only when I moved into New York City, during a billiards boom triggered by Martin Scorsese's *The Color of Money*, that I decided to get serious about it.

In Manhattan, big glamorous pool halls began opening everywhere, which meant I could always find a cheap place to practice. I began by playing a few hours each week, working up to two to three hours every day, until eventually, over the course of a few years, I got darned good. Good enough to see how much further I had to go until I would be truly skilled at the game.

My regular haunt by then was a basement pool hall called Society Billiards, frequented during the day by a handful of hustlers and tournament players who regularly beat me with a humiliating thoroughness. One day, in the course of beating me, a pro named Trevor, who was younger than me but better than I'd ever be, complimented a shot I'd just made. I didn't believe his encomium, it felt like a typical hustle, but I took the opportunity anyway to ask him for some advice about how to get better.

"You're a real good shooter," he said, "your shooting ability is pretty much equivalent to mine. But I just have more knowledge than you. Truthfully, shooting ability is only about twenty percent of what it takes to be a great player. All the rest is knowledge."

Trevor knew table layout, shooting strategies, tips and tricks for getting out of particular jams, and, most importantly, he understood how to play the opponent, not just the table. What he was telling me was that while he knew how to make every shot, as any good pool player does, he more importantly knew how to plan his way through a game *specifically* so that he didn't need to make crazy shots to get himself out of trouble in order to win. That is what made him a *great* pool player. He sketched out every game in his head, prefigured and sidestepped any potential hazards, before he ever took a shot with his hands.

A maker needs to do the same thing with their projects, only on paper. It's not enough to have an idea of what you want to make. It's not even enough to have all the skills required to make it—knowing that you *can* build something isn't the same as knowing *how* you're going to build it. You need to learn how to build a full mental picture of your project, and then you need to draw it out. Rendering it in a physical medium allows you to work out the kinks, come up with an order of operations, refine the details, and also experiment. No matter how complete you might think the construction is in your head, drawing it out will reveal things you had never considered. Like list making, it can feel like a wasted step between ideation and creation, but that's impatience talking. Properly addressing your work, getting things right the first time, and getting them done in a timely fashion often begins with this very step.

In 1986, I made a plaster bandage casting of my torso and it sparked an idea for a sculpture with a heart bursting from a hole

I WANT AN interesting knot tying torso to frame - Fisherman's or something

Shoelaces - many?

Torso hung like tanned hide from natural wood frame

torso is only half shell w/ semi square hole + a heart strung in the hole the same way

heart is only color

torso - white
wood - Natural

Base is a log in half

Back side view

Perhaps torso will have tabs to tie to frame.

I'm quite pleased with the sculpture that came from my sketch: Water of Faces: Torso *(see previous page). It represents, with some slight modifications and embellishments, exactly what I was trying to accomplish creatively. Look at how flat my stomach was back then!*

in the casting. A classic angsty late-teens statement about life and love. Did I mention that the heart was made of razor blades? It was my hipster St. Sebastian, and while I could see the idea in my head, I knew it would take a fair amount of rigging to mount, which would require some fleshing out and some experimentation to make sure it worked both practically and aesthetically. So I sat down and sketched it out. I played with a front view and a rear view. I diagrammed part of the frame to get a sense for the amount of space I would need to secure the casting properly. I made a bunch of notes as I sketched, describing what I was looking at and what my intentions were with each creative choice, so that I could pick up right where I left off in case I had to put the project down for a while, since I was working full-time.

The sketch I made was crude and mediocre (like I said, I've never considered myself very good at drawing), but the sketch's quality is beside the point. You don't have to be a great illustrator to derive the benefits of using drawing as a way to flesh out your idea. What matters is that you faithfully capture the intention of your idea so that once you set your mind to building it, your hands will follow.

DRAWING AS MOMENTUM

Drawing things out is by far the most useful, analog way for me to plan a project. It's also the easiest way I know to keep a project moving forward. Making things, just like physics (of our universe at least), is all about momentum. When you're in the flow, the momentum of a project is hearty and good. When you hit a roadblock—writer's block, builder's block, designer's block—there is no movement and the pathway forward is not at all clear, which often kills momentum and can stall a project. Here's something true: there will be times in *every single build* when a pathway isn't clear. These roadblocks, or momentum killers as I've come to refer to them, aren't outliers. In fact, they are abundant and nobody is immune to them. Financial problems, procrastination, fatigue, family obligations, mistakes, accidents, lack of interest, lack of time, poor feedback—any one of these things can bring your project to a screeching halt and destroy all the creative momentum you've gathered.

I frequently use drawing as a tool or a technique to break through that dam. Drawing always gives me a new vantage point on the project and allows me to see the thing I'm building with enough distance to identify the next step more clearly. In that regard it's almost immaterial *what* I draw. I might draw some reference pictures for a collaborator to understand what I need from their contribution, or to see where their contribution fits in the wider picture. I might draw some mechanical subassemblies that are kicking my ass. I might re-draw the item I'm making for fun, just to stay inside the construction in my head. I might draw a case for the object, or a case I'd like to build for it when it's done. Sometimes the exercise of thinking about what might contain the thing I'm working on can help me define better what it is I'm actually building and help illuminate what has me stuck. It's all information. A conversation between my brain and my hands.

The most intractable momentum killer I face, the one that is most difficult for me to overcome, is confusion. I love the plaster bandage sculpture I made when I was nineteen years old, but thirty-plus years later, that would likely have been the preamble to something far more ambitious like a fully operational Iron Man suit (because that is the most complicated thing I can think to do with a plaster casting of my torso). A highly complex project engages all my skill, all my knowledge, and all my management abilities, but it also introduces confusion to give those creative assets a run for their money.

With so many individual components, so many different materials and methods, so much that has to fit together just right, in just the right order, the more complex a project the more likely it will be to confound you at some point in the process. You can't make this part until that part is finished. You can't assemble these pieces until that one is painted. When you have forty parts to make for a build, it's easy to spend a precious day building something wrong, and man it's hard to work up the energy to build that thing *again*. Drawing, in those moments, has helped me fight off that momentum killer because it invariably increases my understanding of the physical totality of the thing I'm working on. My initial renderings—the drawings that translate the original idea from my brain—help me realize the object I'm creating at a macro level. Drawing as a method to crack through complexity, on the other hand, is to zoom in on the object and get familiar with it at all of its micro levels. Inevitably, the new familiarity that drawing produces is what unlocks the part of the puzzle that has kept me stuck, and provides the movement I need to get the ball rolling again.

Momentum is, of course, all about movement. Experience has taught me that forward movement, however small, is the only way through a blockage like this. It's important to understand, however, that this movement will never be a straight, linear, unbroken path. It's more like an oscillation, a beat, a rhythm even. When I can't make, I draw to find that rhythm again, to create that little bit of forward movement that is the only thing that can slay the dragon of inertia.

They say the pen is mightier than the sword. In my case, the pencil is my most lethal weapon when I'm stuck. It's become one of the key ways in which I keep a project at the top of my mind, and how I mull over its parts and pieces. Whether I'm on deadline, or I'm distracted by other things going on in my life, or I just need to get through this thing in order to move on to something I'm more passionate about, I will unsheathe the pencil, move a delicious new blank piece of paper into the battlefield, and set myself to drawing in the fight to bring my idea to life.

DRAWING AS COMMUNICATION

From Colossal Pictures to *MythBusters*, Jamie and I have worked on hundreds of projects together, possibly into the thousands.

From simple things like making a fake riverbed for a Nike commercial, to complex objects like a wax and nitrous oxide–fueled rocket based on Civil War plans (that FLEW on its first ignition!!), to INSANE builds like strapping ever bigger rockets on Chevy Impalas in three separate episodes over eight years to answer the Darwin Award's most famous myth.*

In many of these projects, our brains ran on exactly parallel tracks. We always had very different approaches to making and problem solving, but once we had a plan we always knew what the objective was and the direction we needed to head to get there. When it came to expressing the ideas themselves, or planning their construction, however, things were not so simple. Because of our contrasting styles, the only way we knew how to translate an idea from one of our minds to the other's was through a process we liked to call . . . arguing. It could be so fraught that Jamie and I used to joke about how excited we were for technology to evolve to the point where we could have little USB ports implanted at the base of our brains and we'd just exchange thumb drives with our ideas on them. With that day still far in the future (for now), we instead resorted to the one discipline we had both always used to explicate our ideas to ourselves: drawing. We turned constantly to the whiteboard, or pencil and paper, to flesh things out and literally get on the same page. In this sense, drawing was more than just a rendering tool, it was the ultimate collaborative communication tool as well.

I've only met one person more enthusiastically insistent about the power of drawing than I am—Gever Tulley. Gever is the author of *50 Dangerous Things (You Should Let Your Children Do)*

---

* The story goes that an Arizona man attached a JATO unit (a rocket engine used to help heavy planes take off) to his car and met his demise when he launched the car into the side of a cliff 125 feet above the ground.

and the founder of the Brightworks School, the Tinkering School, and the kit-making company, Tinkering Labs. Gever is the country's Tinkerer-in-Chief and a huge proponent of visual communication.

"I've been trying for eight years at Brightworks School to introduce visual communication on equal footing with written and spoken communication," he told me as we discussed the threshold to entry for making, and the problems people have getting started and collaborating effectively. "It pains me that we have kids who are twelve years old and they can't make two boxes inside each other. Like they don't have precision of control to nest two boxes to help a diagram."

Gever Tulley's goal isn't to teach everyone how to draw still lifes or landscapes or to become great artists, he just wants them to be able to communicate in visual form to assist in their collaborations. "People will say to me, 'Oh, I'm not a visual communicator,' but everybody is. Everybody's a visual communicator, everybody is a speaker, everybody's a writer, everybody's an artist to varying degrees, and we shouldn't just allow for that needle to be so close to zero that they feel like they aren't that person," Gever said.

His point is so important and it resonates with me so deeply. I am all those other things—a speaker, a writer, an artist—but if I'm not a visual communicator I don't know where I'd be right now in my career as a maker and a storyteller, because I have no idea how Jamie and I would have lasted as cohosts and collaborators over fourteen seasons. Drawing was essential to our partnership.

That said, I'm not always communicating with myself or my collaborators when I draw. Sometimes I'm connecting with peers and enthusiasts who share a passion for a particular topic to commiserate on a project or share what I've found. There's an online forum I frequent, for example, called Replica Prop Forum, where

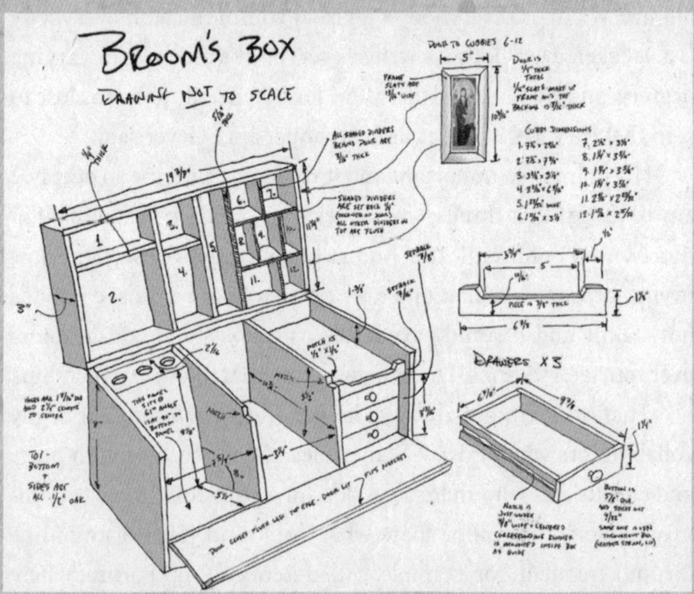

BROOM'S BOX
DRAWING NOT TO SCALE
DOOR TO CUBBIES 6-12
5"
CUBBY DIMENSIONS
DRAWERS X 3
3"

fellow prop collectors, builders, traders, and makers dissect their favorite movie props and artifacts. Some members use the site to chronicle their builds, others use it as a marketplace to buy and sell replicas they've either made or collected. I use it for all of those things.

On one occasion, I bought a *Hellboy* prop on eBay called Broom's Box. Broom's Box is the amphibious Abe Sapien's traveling research box. It's where he looks up esoteric, occult information about whatever foe they might be facing. In *Hellboy* he uses it to describe the demon Sammael to Hellboy and the other members of the BPRD (Bureau for Paranormal Research and Defense). In *Hellboy II*, he uses it to look up the Tooth Fairies released by Prince Nuada that decimate the auction house in which the crown of Bathmora is being sold. Aaaaaaanyway, I was ecstatic about the purchase, and so I took pictures of it and posted it on the forum, along with a list of all the bits and bobs that went in each of its compartments. Almost immediately, I was asked for some measurements of the box itself by members who were interested in making their own versions. So I got drawing and learned so many things about my own prop in the process: the name of the painting on the right-hand door panel; the kind of wood it was made from; where else in the movie you can see the monkey skull.

Since my initial post, at least four people have made their own versions of Broom's Box from my measurements. And years later, my original drawing is still up on the forum, and people are still referencing it for information on making their own. Making things has always been a central pleasure to my life. Drawing has become a late yet equal pleasure, as well, not least of all because it has allowed me to communicate with so many like-minded prop collectors and aspiring makers. It has given me the power to foment their excitement and help expand the boundaries of their own creativity. Drawing can have that same kind of power for a maker of any age or skill level. It is a language in which experts and novices alike can communicate, because it is fundamentally the universal language of creation.

DRAWING AS IDEATION

When I finished my exploding heart through plaster torso sculpture and mounted it to the wall in my tiny studio apartment in Park Slope, Brooklyn, comparing the final product to its sketch inspired me to think about other sculptures I might make along similar lines. Like ritualized three-dimensional versions of Andreas Vesalius's anatomy drawings, I played around in my mind with other body parts that could be well rendered and hung with intention as if on a tanning rack. Eventually, I sketched some of them out.

Like my drawing for the original sculpture, these, too, were mediocre renderings. But just like that original drawing, their quality is not what was important. What was important was that, similar to Broom's Box, drawing the piece I'd made (and various alternatives along with it) helped me learn more about what I'd built. It was exteriorizing all the stuff from the build still bouncing around in my head, in order to gain literal perspective on it. This 2-D recapitulation of my ideas quickly became a fruitful place to look for genuine inspiration and has remained a critical part of my practice as a maker.

You don't have to be great with a pencil for this to work. Like I've said, I've never considered myself particularly good at drawing. For the longest time it never felt like the line did what I wanted it to do, yet I continued to draw. One, because it continued to be useful, and two, because it clearly helped me get better at communicating my ideas more precisely. This is Gever Tulley's entire

Brainstorming sketches for my Water of Faces series; I made almost none of these, but that wasn't the point.

goal with respect to visual communication skills. "The pushback I often get is, 'I don't know how to draw,' and my response is 'Well, how about you go home and spend the summer drawing every day and then we'll talk about it in the fall when you show me your notebook,'" Gever said, rightfully indignant. "Because we know that practice can move your mark making over to something more precise and controllable."

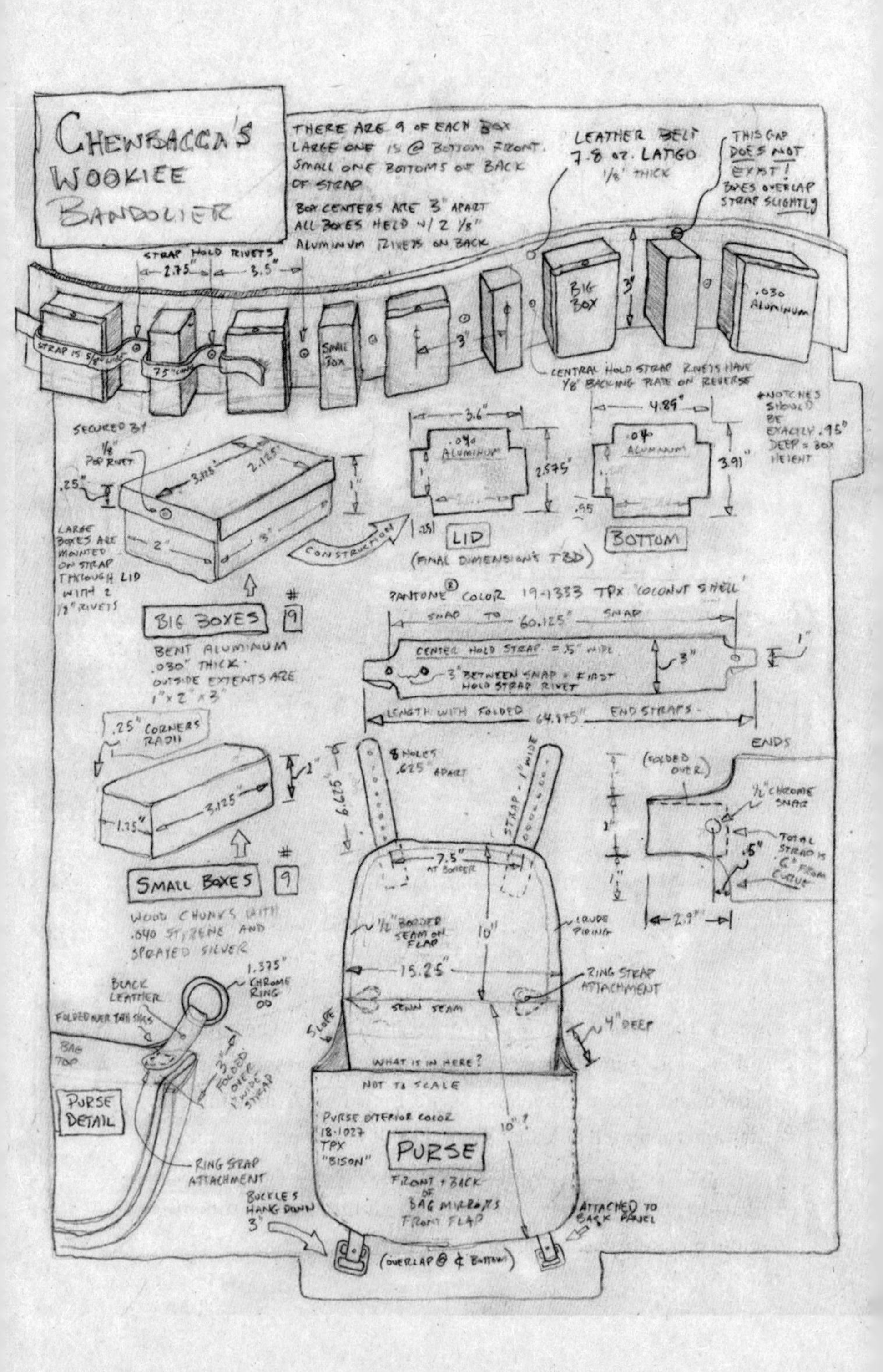
CHEWBACCA'S WOOKIEE BANDOLIER
THERE ARE 9 OF EACH BOX
LARGE ONE IS @ BOTTOM FRONT.
SMALL ONE BOTTOMS OF BACK OF STRAP
BOX CENTERS ARE 3" APART
ALL BOXES HELD W/ 2 1/8" ALUMINUM RIVETS ON BACK
LEATHER BELT
7-8 OZ. LATIGO
1/8" THICK
THIS GAP DOES NOT EXIST!
BOXES OVERLAP STRAP SLIGHTLY
STRAP HOLD RIVETS
2.75"
3.5"
STRAP IS 5/8" WIDE
.75" LONG
SMALL BOX
BIG BOX
.030 ALUMINUM
CENTRAL HOLD STRAP RIVETS HAVE 1/8" BACKING PLATE ON REVERSE
SECURED BY 1/8" POP RIVET
LARGE BOXES ARE MOUNTED ON STRAP THROUGH LID WITH 2 1/8" RIVETS
CONSTRUCTION
BIG BOXES
9
BENT ALUMINUM .030" THICK. OUTSIDE EXTENTS ARE 1" x 2" x 3"
.040 ALUMINUM
LID
(FINAL DIMENSIONS TBD)
.04 ALUMINUM
BOTTOM
*NOTCHES SHOULD BE EXACTLY .95" DEEP = BOX HEIGHT
PANTONE® COLOR 19-1333 TPX "COCONUT SHELL"
SNAP TO 60.125" SNAP
CENTER HOLD STRAP = .5" WIDE
3" BETWEEN SNAP + FIRST HOLD STRAP RIVET
LENGTH WITH FOLDED 64.875" END STRAPS
.25" CORNERS RADII
SMALL BOXES
9
WOOD CHUNKS WITH .040 STYRENE AND SPRAYED SILVER
8 HOLES .625" APART
STRAP - 1" WIDE
7.5" AT BORDER
1/2" BORDER SEAM ON FLAP
CRUDE PIPING
15.25"
RING STRAP ATTACHMENT
SEWN SEAM
4" DEEP
SLOPE
WHAT IS IN HERE?
NOT TO SCALE
PURSE EXTERIOR COLOR 18-1027 TPX "BISON"
PURSE
FRONT + BACK OF BAG MIRRORS FRONT FLAP
ATTACHED TO BACK PANEL
BUCKLES HANG DOWN 3"
(OVERLAP @ ₵ BOTTOM)
ENDS
(FOLDED OVER)
1/2" CHROME SNAP
TOTAL STRAP IS 6" FROM CURVE
2.9"
BLACK LEATHER
1.375" CHROME RING OD
FOLDED OVER TAB SEWS
BAG TOP
3" FOLDED OVER 1" WIDE STRAP
PURSE DETAIL
RING STRAP ATTACHMENT

And if that isn't enough, you can always find help in other places. I draw inspiration from the drawings of others. I never tire of poring over the drawings and graphic novels of Moebius, for instance. I get a lot out of looking at Ridley Scott's storyboards (he's a wonderful draftsman). Since I was a kid I loved all those old drawings from the mid-century issues of *Popular Mechanics*. Something about their clean lines and multidimensionality and the way the artists kept all the pieces separate yet constantly oriented to each other, spoke directly to how my brain looks at ideas. I've extracted more than a few ideas directly from the line drawings of that great magazine. Later in life, they also inspired me to draw out something I'd built several times before, to get its proportions exactly right, so that I could build it again inside a day any time I wanted. I'm talking about Chewbacca's bandolier.

Chewbacca is the greatest nonhuman character in film. I am so in love with Chewbacca that I've built his bandolier a half-dozen times, partly because it's just fun and partly because every time I find new reference material (which happens more than you'd think) that by definition necessitates a new construction. Most recently, I was able to solidify some of my measurements based on a couple key pieces of data I received from a source deep within the *Star Wars* universe. These dimensions TOTALLY changed my perception of the bandolier and forced me to reckon with the fact that I'd done a LOT in my previous five or six builds that was way off the mark. The fact was, I'd been using reference material that spec'd the silver ammo boxes as nearly twice as big as they should be. Whoops!

Now that I had the true, canonical measurements, I didn't just have to remake the bandolier, I HAD to memorialize it, and making a simple list of the correct specs wasn't enough. It wouldn't communicate the unique complexities of such an object. I decided that the best method was to illustrate as best I could everything I now knew about making the boxes. I started with the wavy bandolier on the top and worked my way down, capturing every critical measurement and spacial relationship I'd assembled for its components over the years. I could make a new version of Chewbacca's bandolier tomorrow based on this drawing. So could anyone I give the drawing to. That fact alone makes me so happy, and keeps both the inspiration engine and the permission machine firing on all cylinders.

Like I said before, I draw every day. Often to work out ideas on paper, to keep momentum going on things that are too complex for me to hold in perfect orientation in my head, sometimes to codify measurements that I'm working through, or to communicate with those helping me turn my idea into reality. Mostly, though, I love drawing things I have made and things I might make in the future, because it makes them real to me in an immutable way that feels deeper than just an idea or a memory.

Cellist David Darling once recommended that musicians should seek to "sing everything they can play and play everything they can sing." I think that making and drawing occupy the same spaces as singing and playing. They use different parts of your brain as individual pursuits, but when combined they become symphonic.

INCREASE YOUR LOOSE TOLERANCE

I have a prediction: you are going to mess up a lot. I mean A LOT. Whether from impatience or arrogance, inexperience or insecurity, lack of knowledge or lack of interest, you are going to tear seams, break bits, snap joints, misdrill, overcut, under-measure, miss deadlines, injure yourself, and generally just make a mess of things. There will be moments when, if you are not losing interest in a project, you are losing your mind about it. It will be confusing, dispiriting, and infuriating.

About this prediction, I have three words for you: WELCOME TO MAKING!

It's an exciting time, for we live in a truly golden age for creativity. If you desire to pick up a new skill or learn a process, or even an entire discipline, there's almost surely someone online who's helpfully filmed a video about it. Three hundred hours of video get uploaded to YouTube every minute and a significant percentage of those hours are devoted to showing people how to make things. From bowl turning, to welding, to scuba diving, to guitar building, to animal husbandry, you can learn anything online. I would have killed for such a resource early in my career. What a gold mine of generosity and information, all these people everywhere willing to share their knowledge and experience. There is one bit of wisdom most of these people do not share, however. There is often something missing, in my opinion, from many of their videos: how fraught the process of making can be.

Making is messy. It's full of fits and starts, wrong turns, and good ideas gone bad. New methods, new skills, new creations, they are all a product of experimentation; and what is an experiment but a process that may or may not yield expected results? WHO KNOWS?

In hotbeds of innovation like Silicon Valley and Seattle and Austin (each of which has amazing maker spaces), when people talk about this aspect of creativity, they like to use the word "failure." They bake it into lots of catchy phrases: "Fail fast," "Move fast and break things," "Learn to fail or fail to learn." And we lap this rhetoric up. The (mostly white male) billionaire entrepreneurs who evangelize for failure have dominated our cultural landscape for the first twenty years of the twenty-first century to such an extent that we have fetishized them and their ideas. Living in San Francisco, I cannot begin to tell you how often I hear that we need *more than anything* for our kids to safely "learn how to fail."

I admit, the word "failure" is catchy, it grabs your attention. "Fail" is one hell of a four-letter f-word. It's sticky, in marketing parlance. It's useful because it is universal. We have all failed in our lives and we will continue to fail. It's called being human. Failing at things, genuinely screwing up, is an inescapable part of the human condition. I frankly don't trust people who say they haven't failed at anything. Facing yourself, and the places in which you have not performed to the degree you'd hoped lends a vital tonic to one's character.

But in the context of creativity what we're talking about here is not truly failure per se. True failure is dark, it hurts, it affects others, and it's something to be recovered from. Failure is getting drunk and not showing up to your kid's birthday party. None of that describes the type of failure that's become such a buzzword.

In all this talk of failure, what we are really talking about is iteration, experimentation. What we are talking about is the freedom and willingness to try a bunch of new things in the pursuit of new ideas until we find the thing that works. Because creativity does not move in a straight, unbroken line. It is a path that twists and turns and doubles back on itself sometimes. It's never linear. There will be "wrong turns" that feel right and seem to be taking you in the general direction of your goal, but gradually veer far enough off course that you have to backtrack to the fork in the road and take the other spur—the other branch on the decision tree.

On one build, I was painting a burnt, rotting robotic hand I'd made to look as if it were three hundred years old, and I could not for the life of me figure out what that paint scheme should be. I tried a rusty, greasy finish—exactly the finish you'd expect to see on a futuristic-yet-rotting robot hand—and to my surprise it didn't work at all. At that point I was out of ideas. This sculpture clearly needed a decidedly unique finish. What it should be, however, I had no idea. All I knew was that it needed to be something radically different. With no sense for what was right, as an exercise I asked myself if I knew what was *wrong*. What would be the *worst* possible finish for a robot hand from the distant, fictional past?

To my surprise, I actually had a strong opinion about that: candy cane. Barber pole. Red and white stripes winding around the piece. That would be *perfectly* awful. So I spent the next five hours meticulously masking and painting this elaborate hand to be candy-striped red and white. And, of course, I was correct in my assumption: it looked horrible! Most importantly though, the *second* I pulled off the masking tape I knew what it needed to be. It absolutely needed to be finished in a verdant, leafy green, a finish I executed and still love to this day.

Some branches you follow will require you to push yourself far enough in that direction to know it's the wrong one. It's all part of the process. Creation is iteration. Your job as a creator is to take as many of the wrong turns as necessary, without giving up hope, until you find the path that leads to your destination.

It doesn't matter if you're a model maker, a potter, a dancer, a programmer, a writer, a political activist, a teacher, a musician, a milliner, whatever. It's all the same. Making is making, and none of it is failure. It is an iterative process. It is how you learn new skills. It is how you gain knowledge and experience. It's how you improve yourself. It is how you make new things. And the key to it all is tolerance. Both literal and figurative.

To be a successful maker you need to have significant patience and endurance, but you also need to open your mechanical tolerance. In engineering, tolerance refers to the amount of permissible variation in the manufacture of a physical object. If you order a bushing from a machine shop, you'll often say you want it built to a specific measurement, with a tolerance plus or minus five thousandths of an inch. In *mechanical* engineering, where I spend most of my time, it refers to the amount of space between something like a bolt and a nut into which it is designed to thread.

Imagine hanging a door and sliding the steel pins through the holes of each door hinge. If the pins fit loosely in the holes, and rattle around a bit after insertion, that is a loose tolerance. If the pins fit snuggly, and need a tiny bit of pressure to thread through the hinges, that's close tolerance.

There are advantages and disadvantages to various tolerance levels. How close your tolerance needs to be dictates a lot about the limits of how you should proceed in building and using something. If you're joining two plates with four bolts, and you want a close tolerance fit for the bolts that will live in those holes, you better be darned exact about your hole placement, possibly even use a precise milling machine to tap the holes. But if you only need a loose tolerance, drilling slightly oversized holes for the bolts allows for small variances to be . . . wait for it . . . *tolerated*, and thus the hole pattern could be drilled on a less exact drill press.

Close tolerance yields efficiency from beautiful parts that fit together incredibly well, though that accuracy comes at a cost. It's time-consuming and the equipment you're working with requires a high level of maintenance and calibration to achieve repeatable accuracy and smooth finishes.

Loose tolerance, on the other hand, provides durability from parts that can take a lot more abuse because dents, dings, and dirt have room to move without inhibiting the performance of the larger machine. The Kalashnikov rifle is a great example of loose-tolerance engineering yielding high reliability in the field. The AK-47 was designed with loose tolerance all around so that it could be repaired, rebuilt, and modified to one's heart's content in the dirtiest conditions imaginable. It is no coincidence that it is the favorite of guerilla fighters and armies from poor countries the world around.

Close and loose tolerance can also be the difference between expensive and cheap. One of the main differences between a cheap car engine and an expensive one is tolerance. High-end car engine parts might be machined with an accuracy to the ten-thousandth of an inch, while a cheap one might be machined to around a thousandth of an inch. This sounds like the very definition of splitting hairs, but in fact the difference is marked. When a pin fits more loosely in its hole, it can rattle and vibrate. Vibration is an energy vampire, stealing energy, and thus performance, away from the intended movement of the mechanism to which the pin is a part. Making the fit between pin and hole tighter allows for a more efficient transfer of energy.

This concept of tolerance will be very important as you learn not just to engineer more complex objects, but to actually complete them. For it is essential that you build a loose tolerance into your process, to give yourself room to mess up. Something I call "mistake tolerance"—a term I'm inventing here and now. In the past, so many of us have been afraid of this idea. We were terrified of screwing up, because if we didn't "get it right" then it was a waste. There was no tolerance for waste—of time, money, talent, other people's patience. Except this is precisely where you figure out what something is supposed to be and how best to make it. If you don't give yourself enough room to maneuver and to mess up, neither may happen.

LEARN BY DOING (POORLY)

I often describe myself as a serial skill collector. I've had so many different jobs over my lifetime—from paperboy to projectionist, from graphic designer to toy designer to furniture designer to special effects model maker—that my virtual tool chest is overflowing. Still I love learning new ways of thinking and organizing, new techniques, new ways of solving old problems. But I'm never interested in learning a new skill for its own sake. And the skill itself is rarely the thing I care most about. It's often a bonus side effect of my obsession with making, or my desire for having, some *thing.* The skills I have, all of them, are simply arrows in my mental quiver, tools in my problem-solving tool chest, to achieve *that* thing. They are each of them only a means to an end. And I learned each of them specifically for that reason. Eventually, and I don't know if I was always like this or if it developed with time, I came to realize this was the ONLY way I could successfully learn a skill—by doing something with it, by applying it in my real world.

One of the very first "skills" that I picked up was juggling. It was my dad's go-to when they entertained or whenever there were three apples conveniently nearby. I desperately wanted to be like my dad, so of course I wanted to juggle, too. Unfortunately, as someone who was never athletically inclined, tossing and catching multiple balls without dropping them was a feat of hand-eye coordination that I couldn't seem to manage. But after combining countless weeks of dropped balls and muffled swear words in my bedroom, along with a book called *Juggling for the Complete Klutz*, I unlocked the latent circus performer inside me.*

When I was twelve or thirteen, I became interested in model trains and dedicated my bottom bunk to a pretty detailed set. From a handful of library books about model-train making I was inspired to create a bigger world around the set. This is a significant part of the model-train hobby for true ferroequinologists (a word I also learned in those library books). Model-railway enthusiasts love to make their *own* stuff: landscapes, buildings, entire towns. Not kits, either, but working from raw materials in what is called scratch building. I'd built plenty of plastic model kits but scratch building felt like adult-level modeling to me. With the help of those books and my dad, I learned how to make a two-story building out of matte board, I learned how to mix

* I'd augment this circus skill when I was sixteen years old by spending the summer learning how to ride a unicycle.

color for painting the roof tiles, and I learned how to use planar forms to create compound curves for the rotunda we built in the middle of the fictional town that quickly grew up around the train set.

In my mid-twenties, in between theater gigs in San Francisco, I worked as an assistant for a machine artist and robot maker named Chico MacMurtrie at a place called Amorphic Robot Works. Chico built incredible robots that danced, played drums, procreated, became buildings, and told stories. I learned a ton in Chico's shop, but one thing I will always remember about that time is that his shop was the first place I encountered a lathe.

Chico had obtained this beautiful old 36" Craftsman lathe. He walked me through its basics: how you tighten stuff in the chuck, how to turn it on and off, and how to put tools into the cross-feed in order to cut into the work that was in the chuck. It was very cool to look at, but my interest in *using* the lathe didn't develop until I had something I needed to make, for which the lathe was uniquely suited.

San Francisco in the early '90s was one of the great garage sale towns in the United States. Every weekend you could find hundreds of people clearing out their cupboards and closets, moving, and selling old stuff on the street. We called it garage sailing. It was a regular weekend occupation of mine, one that I brought with me from my earlier "found object art" days in Brooklyn, but that evolved here on the West Coast thanks in large part to finally having a job and a little money to spend. One weekend making the rounds, I picked up a beautiful, ornate, folding traveling chess set made out of wood. It was probably seventy-five years old and its craftsmanship was incredible, but it only had half the pieces. It had white, but no black pieces, so I decided that I would make them on Chico's lathe. The fact that I'd yet to log any meaningful experience on the machine was beside the point. The chess pieces would give me some meaningful experience.

This is where the importance of a loose tolerance with yourself comes into play. Back at Chico's shop, I described to my friend Geo what I wanted to do for the chess set and we got into a philosophical argument about how to use the lathe. Geo was a master maker in his own right, but his approach was radically different than mine. Geo was a computer scientist, and a superb engineer, methodical and sober in his approach to everything. Geo could grab a random screw off the ground and tell you if it was a 10-32 fine-threaded or a 10-24 coarse-threaded machine bolt just by looking at it. He was all about accuracy. Geo insisted that we proceed by drawing out side elevations of each piece we wanted to cut, make measurements of those drawings, then transfer those measurements into a set of cuts on the lathe to achieve the pieces we wanted.

I was more like a train rolling down the tracks without a brakeman, and my approach was equally straightforward: I wanted to just start cutting metal until it looked right. This confounded Geo. We'd be wasting so much time and material! I recognized his sense of exasperation and true confusion when Jamie and I joined forces for *MythBusters.* Jamie would often joke that if each of us had four hours to do a project, he'd spend three and a half of those hours drawing and planning out everything he needed to build something once, whereas I would spend almost no time planning, then try five different solutions until I found one that worked, and we'd both come to the same result in the same amount of time.

With Chico's lathe, I wasn't interested in learning how chess pieces are made. I wanted to use the lathe so I could *have* the chess pieces I needed to complete my set. Diving right in was how I was planning to learn the feel of the metal lathe. Turning the crank to start a tool cutting into a spinning aluminum billet that would eventually become a chess piece gave me a physical feel for the tool, and for the metal I was manipulating with that tool. Mistake by mistake, success by success, I learned more about moving metal and the rhythm of lathework that afternoon than any amount of reading would have taught me. I'm not saying that empirical knowledge and intuition replace reading, I'm saying they augment it. They fulfill an area of knowledge that reading can't reach. *Doing* puts the kind of knowledge in your body that can only be gained through an iterative process.

Every maker needs to give themselves the space to screw up in the pursuit of perfecting a new skill or in learning something they've never tried before. Screwing up IS learning. One of the best ways to do that is by giving yourself a cushion with materials. If you're an aspiring fashion designer, for example, and you have a new dress idea that calls for four yards of fabric and some unique construction methods, when you go down to the fabric store don't buy four yards of fabric. Buy eight. Buy TWELVE. If cost is an issue, buy half of it in the cheapest fabric you can find and half in the expensive stuff you actually want. This way, when you cut a piece wrong or rip through a seam or spill a cup of coffee on the skirt (all of which will happen at some point, believe me), you've done it on the cheap stuff. Then once you get the rhythm and shape and the pattern right, you can use the cheap version as a template for cutting out the dress pieces from the more expensive materials and you haven't wasted a thread of the good stuff.

This approach applies to any creative discipline. If Traci Des Jardins is cooking a three-course meal for twenty people and the menu includes Cornish game hen and soufflé, she's not going to buy twenty game hens and just enough ingredients for twenty soufflés, especially since she's not a great baker. She's going to buy twenty-five hens in case one of them falls on the floor, or she cuts herself during prep and blood contaminates one of the birds. And she's going to bring enough ingredients for twenty-five soufflés, because one or two of the delicate desserts will inevitably fall. Those extra ingredients give her a cushion that reduces the stress of cooking a fancy meal for lots of people and provides space to try new things in the moment.

I do the same thing any time I embark on a complex build with materials or methods I don't have a long track record using. If I want one thing, I start by making three. If I want five things, I plan to make eight, because invariably I'll screw up along the way, and having extra parts is my insurance. It's a buffer for (inevitable) mistakes. It's my relief valve. My plan always is: I don't know how circumstances will change, only that they WILL change, so I'd better be ready. When I built the Mecha Glove from *Hellboy*, I made four of every component and commissioned four of every

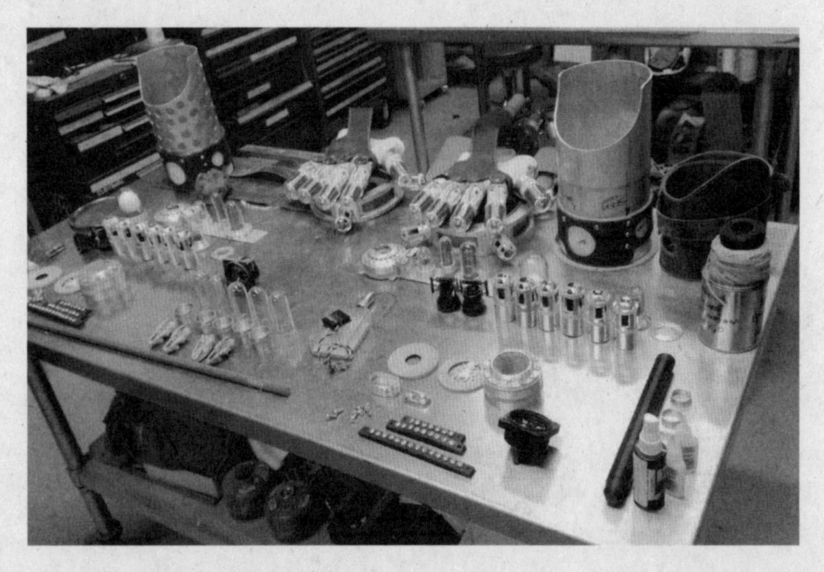

I actually began this project with a drawing. I decided to do it on the spur of the moment and didn't have a protractor with me so I made my own out of a coat hanger.

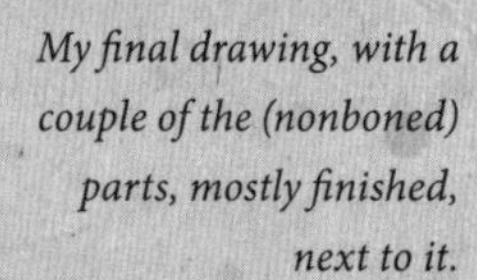

My final drawing, with a couple of the (nonboned) parts, mostly finished, next to it.

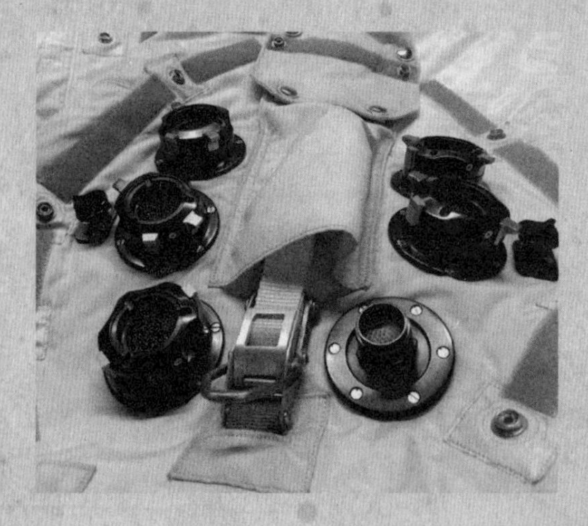

Once all the parts were completed, I sent them to an anodizer to add the distinctive red and blue finish of NASA hardware. I can't tell you how exciting it was to get them back, it was like Christmas!

glass tube that encased a finger. By the time I was done making mistakes or going down wrong paths, I had enough for two—one for me and one for Guillermo del Toro.

Late in 2017, I wanted to bring my machining game up a level, so I decided to get there by machining something really difficult: the couplings on the front of my Apollo suit replica. There are six connectors on the front of an Apollo suit chest, only two of them identical, which meant I had to make five very similar yet subtly different parts, each requiring eleven different machining setups—a level of machining complexity I had not yet explored, which presented *plenty* of opportunity to screw things up. I accommodated for the potential screwups by gathering enough material to make eight connectors, just to be safe. It was just enough. I completely boned two of the parts during the early stages of the manufacturing process. Fortunately, having learned from those two early miscues, the others came out fantastically and ultimately so did the suit.

CREATION IS ITERATION

Gever Tulley told me an amazing story from the early days of Brightworks, the maker-based charter school he founded in an old mayonnaise factory in the Mission District. It was maybe the first or second year the school was open, and one Monday morning one of the teachers removed all the chairs from her classroom. When the students arrived that day, there was much confusion. "It was like, 'Oh no, what happened to our chairs?' And these kids were our seven-year-old students," he remembered.

SEVEN! They were basically first- and second-graders, and their brilliant, gutsy teacher was giving them a choice: you can stand all year, or you can come with me into the shop and we can make our own chairs. I'm so thrilled by this setup, I'll let Gever take it from here:

"So, they head to our little shop at Brightworks, and they build chairs using every naive assumption they have about things like screws and glue. Some of the chairs don't even make it from the shop back to the classroom. None of them lasted more than a couple of days, they all just tore themselves apart. But every time one broke, its maker picked it up, set it on the table, and everyone sort of scrutinized it like, 'Okay, so what happened to this chair?'

"Eventually, they get back to the place where the class has nothing to sit on again. So this teacher says, 'Let's take another stab at this chair thing,' and they head back to the shop to do a second generation of chairs. This time, having looked at some chairs and seen things like crossbars and various other supports, the kids are being much more precise, because one of the things they discovered is that if the legs are different lengths, the chair tears itself apart.

"Now these second-generation things, they're much better chairs, but just because they're better and they hold together doesn't mean they're comfortable. So the students go back and do a third generation and this time they've really started to look at a taxonomy of chairs. They've been out in the field, they've gone to a furniture shop that makes furniture and sat there and watched as the craftspeople put a chair together with dowels. Then they got a lesson and all these things became clear to them: oh, these have to be cut to one thirty-second of an inch. Oh and this wood the chair is made out of, it's not soft pine, it's some kind of hardwood.

"They take that information back to the Brightworks shop and start to produce this third generation of chairs and instead of being thrown together in a day, like the first generation, instead of being a one-day build, they're figuring out an order to the build so they can get everything glued and compressed. They're planning overnight clamping operations. They're careful to assemble in the proper order because now they know if you do it in the wrong order, you can't actually put the chair together. They've had to think through and do test fits and get everything settled, and then at the end they have a piece of furniture, not a single part of which they don't understand why it is the way it is."

That Brightworks teacher brought a group of seven-year-olds through an iterative process that produced complete, thought-out, fined-tuned creations that were unique to each student. She gave them room to try new stuff they saw in other chairs, to make mistakes, and, more importantly, she gave them space to correct their mistakes in each subsequent iteration of chair. At the end of the year, every kid got to take their chair home. They'll likely have them for the rest of their lives, along with the bevy of skills they learned in the process. These chairs are a physical manifestation of a story about iterative learning.

I took a very similar approach when my own son, Thing1, wanted to make a belt holster like mine for his multitool. For Christmas that year he'd been given a Leatherman multitool (the

EOD bomb squad Leatherman, if you must know) and wanted to make an aluminum holster for it similar to the one I had. Of course, my holster was not something I "had," it was something I *made*. Something I had built for myself numerous times.

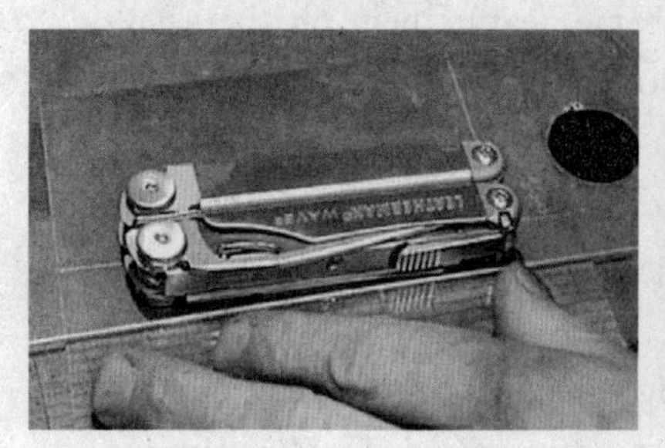

I've worn a multitool on my belt for thirty years. If a spider's web is truly, as some animal researchers assert, a manufactured extension of the spider's nervous system, then the multitool is my third hand. I use it so often that when it's not on my belt I end up with "phantom multitool" syndrome, where I am sure I can feel it there on my hip even though I know it's not there. My current sidearm is a Leatherman Wave. When I was a full-time model maker at ILM, I found the Leatherman so consistently useful that I got tired of the snap-top leather sheath it sat in—it took too long, in my opinion, to unsnap and resnap every time I needed the tool (a small glimpse at how impatient I was/am), which was approximately fifty times per day. So I designed and made my own sheath out of aluminum sheet and a strip of leather.

I spent a few hours putting the first prototype together, including messing up a few times, before I landed on the form that I wore on my belt for the better part of ten years. Eventually, my metal fabrication skills advanced enough that I could make the whole thing out of a single piece of aluminum and form it in such a way that the tool's trapezoidal shape fit perfectly inside it and only needed one finger to release. I could hang upside down with this thing on my belt (which I have done) and never worry about the tool falling out (which it didn't). So I rebuilt the holster, and then rebuilt it again, perfecting it with each iteration. The version I currently wear is the best one yet. It even makes such a satisfying *ker-CHUNK* sound when you take out the tool or put it back, that I've had sound designers record it for their libraries of random mechanical sounds.

My son's no fool, he knew I made my holster, and he wanted to build his own. When he asked if he could come to The Cave (my current shop which, like Brightworks, is also in the Mission) to make it, I said absolutely! But I explained he should be completely prepared to build it at least three times. He sighed, dejected, until I told him what I just told you: my first holster took four tries and the one I was currently wearing took two. He courageously started and indeed went through exactly three versions before he built the holster he wanted. And just like those young Brightworks kids, he still has his creation and uses it to this day.

Mistake tolerance is particularly valuable in this aspect of the creative process. When you know what you want to make, but you're not exactly sure what it should look like or how it should operate, you need to give yourself permission to experiment, to iterate your way there. That's not just how you get to what you want, it's how you get good at it. You have to do it over and over and over again. Anticipating mistakes is how you put space around the unfamiliar and the unknown. My son was bummed out by that concept at first because he inherited my impatience and wanted to be able to make his holster once and be done with it, but if you expect to nail it the first go-round every time you build something new—or worse, you demand it of yourself and you punish yourself when you come up short—you will never be happy with what you make and making will never make you happy.

KNOW YOUR TOLERANCE

One of the great benefits of allowing yourself the room to experiment and make mistakes is that over time you develop an instinct for when a build demands close tolerance or loose tolerance, both in the actual construction of it and in the mental approach to it. This is what Andrew Stanton was displaying with that project he consulted on at Pixar. In telling the group their early attempts were going to suck, he was pointing to the loose tolerance baked into that particular project and he was giving the group permission to get a little messy with it, in kind. He was giving them room to iterate. It's a skill he refined over time, but that originated at his place of work.

"The thing that Pixar learned early on and institutionalized was what if you got it wrong? How do you handle it?" Andrew told me. He was talking about how he thinks of the way Pixar develops films. He compared it to digging up dinosaur bones. In the beginning, you have an idea, and you think you know how things might go, and you make your best guesses, but really you're mostly bullshitting. "It's no different than going, 'Based on my instincts, in this plot of ground I think there's a *Tyrannosaurus Rex* deep down here,' and then you start digging." When it goes right, though, "it's like the only thing you can take credit for is that you picked the right plot of dirt to start digging in." But it rarely goes so swimmingly. Here's how Andrew described that moment inside Pixar when they realize they've gone down the wrong path:

"So what if after you dig all these bones up, and usually by then you're three years into a four-year project, and you put all the bones together and suddenly you go, 'Holy shit! This tail bone is actually the neck bone and this neck bone's actually a tail bone . . . and, wait, I actually dug up a *Stegosaurus*.' Are you going to have the intestinal fortitude to admit to your crew, to your bankers, to everybody after all these years of promising a *T-Rex* that you actually have a *Stegosaurus*, because that's true to what you found and uncovered? We do that. We're no better at telling stories and making stories; we just have the guts to admit what we're really digging up."

Pixar teams have had so much experience with the iterative nature of storytelling and filmmaking that they've built enough tolerance into their own process to not only allow for these missteps and wrong turns, but to have a system in place for responding to them. It takes a lot of intestinal fortitude, as Andrew calls it, but it's hard to argue with the results. In the end, they've made some of the most affecting films of the last fifty years, animated or otherwise, many of which have grossed into the billions.

Not all projects can afford loose tolerances, of course. Some can demand incredibly close tolerance, for a multitude of reasons. In one of the early seasons of *MythBusters*, we decided to tackle the very famous urban legend about a scuba diver sucked up by a firefighting helicopter. These helicopters dip huge proboscises into nearby lakes and suck up water at two thousand gallons per minute into a large tank, which they then dump onto a fire. Supposedly, a scuba diver was found charred at the top of a tree after a forest fire and the only answer investigators could come up with was that he had been diving in a lake, then was unwittingly sucked up by the copter and dumped onto a fire.

Now these helicopters certainly exist, and so do the water pumps that they use, but nobody, and I mean nobody, would let us use one for the show. We spent weeks calling all over the country just to continue hearing "no" from every fire department, forest service, and county sheriff who would take our call. Our reputation often got us cool access to stuff, but in this case it worked against us using these very expensive pumps for our show. Nobody wanted to take the chance with such an important piece of equipment.

When we finally realized we'd never get the pump, my producer turned to me and said, "I guess you'll just have to build it." There were a couple difficulties with this: one was that I'd be mostly on my own because Jamie was incredibly sick with a flu, and was out of commission for two weeks. If you knew Jamie, for him to stay home for two whole weeks is a world-shattering event. Even he was impressed with the bug that kicked his ass. The second problem was that I had never built a machine that big before. I understood exactly how it ought to work, but almost all the machines I had built up to that point were more like the size of a toaster oven, or maybe a person. This one needed to be about fifteen feet tall and it had to work absolutely perfectly. Because this was a build that would aid in testing our story, there wasn't an option for it not to work. Failure, in this case, wasn't an option. There was no margin for error.

Why? Well our design for the pump called for an eight-foot-long steel input tube, twelve inches in diameter, at the top of which was an angle-cut and welded piece of the same pipe set as the output. Mounted above the pipe would be a 250-horsepower outboard motor that we found on Craigslist. The eight-foot input pipe required the shaft of the motor be extended an equal distance down the center of the pipe with the motor's propeller mounted at the bottom, drawing the water up into the long pipe with what we hoped was enough force, and then spitting it out the angled top.

I began by simply wondering what will the failure modes for this object be? When you're extending the output of a large, fast, powerful motor, your first enemy is that insidious energy vampire, vibration. You can't just weld a pipe to it. That pipe needs to be

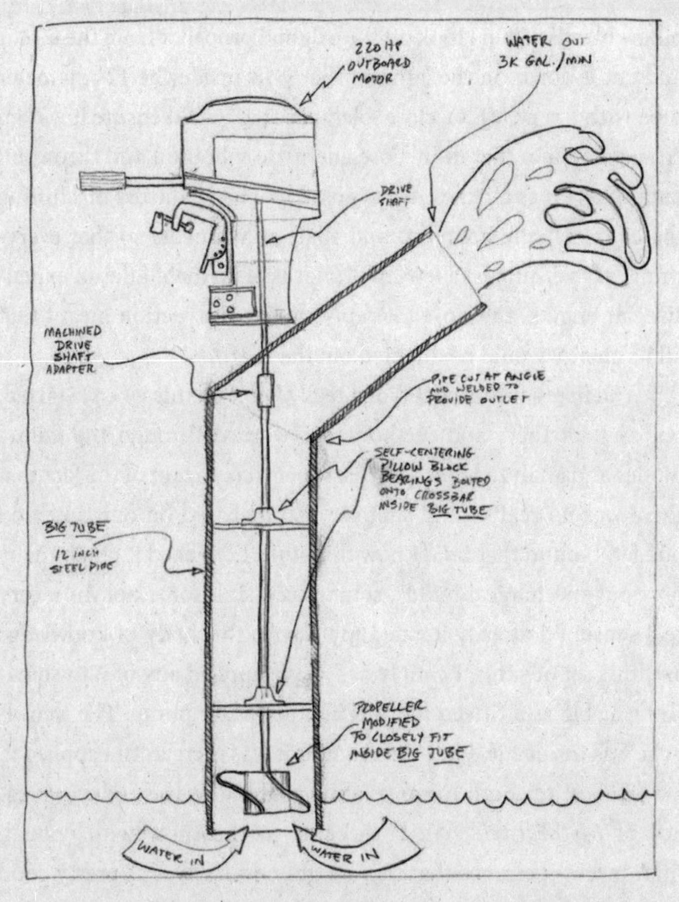

Trust me when I tell you that this pump churned up water at a fantastic rate, and with an incredibly high tolerance to boot.

highly concentric and stabilized so that it spins true. You can't simply stabilize it top and bottom, either, because at high speeds even rigid hardened steel shafts can bow outwards from the centrifugal force (it's called throwout). So I designed an internal system that would hold the output shaft of the motor at several points along its length with waterproof, high-speed, self-adjusting bearings called pillow blocks. Then I took off the original propeller from the motor and cut it down on the lathe so that it fit inside the 12" diameter tube with a very, VERY close tolerance that would ensure it was an efficient system that didn't lose energy to vibration and throwout, but pulled up as much water as possible. Then I started machining the couplers, and adapters, and shaft attachments so that everything was within just a few thousandths of an inch. Like an expensive car engine, the close tolerance of its construction meant that all its energy would be directed out the shaft.

And here's the thing: I didn't really *know* all this when I started, but as I sat there and methodically worked through the failure modes and analyzed how to bypass them, cognizant of the fact that there was no real second shot at this one based on our time and budget, I found that I *did* know this stuff. I'd picked it up on many previous mechanical builds, some successful, some not. In a very real sense, I'd already iterated my way to this body of knowledge and this set of skills, I simply had never applied any of it to something as big as a fifteen-foot firefighting water pump. The knowledge was inside me, I just didn't know it was there until I applied it.

This is, for both me and Jamie personally, the enduring legacy of *MythBusters*. We both came to the show with enough high-level experience that we thought of ourselves as pretty good engineers and problem solvers, but looking back from the other side of fourteen years on the show together, we can both see so clearly what innocents we were when we started, and how much both of us learned by giving each other permission to push past our perceived limitations and giving each other the space to screw up as we did it. The results speak for themselves frankly, both in the quality of the show and in our respective skill bases, which increased tenfold by the time we walked away. Both of us were changed fundamentally by the experience.

It was quite a feeling when we fired up the pump that first time and watched it work its magic, particularly with the limitations under which we operated on that episode. In the end, Jamie recovered in time to help me put the final touches on the build, and when we turned it on it didn't just pump the two thousand gallons per minute we were hoping for, it pumped THREE THOUSAND GALLONS PER MINUTE.

I had another experience with a close tolerance project the following year when *MythBusters* entered its third season a bona fide hit. I was making good money, and I could finally afford to make a perfect *Blade Runner* blaster replica—my third, but this time from the real gun parts! My previous versions were solid yet mistake-riddled attempts that reflected where I was in my life and skill level at the time I made them. The first was somewhat cartoonish, made with parts from a toy gun I found in a shop on Canal Street in Manhattan in the late '80s. My initial sketches for that build came from watching a *Blade Runner* VHS tape over and over again on a crappy 19" TV/VCR combo unit. The second, I made while a full-time model maker at ILM in the late '90s. Thanks to some reference material I found in a hobby shop in San Francisco, I was able to sculpt it to a great degree of accuracy. It was just 20 percent smaller than the actual prop.

Finally in 2005, I was able to do it right, in no small part because back in the '90s a couple of avid *Blade Runner* fans, Phil Steinschneider and Richard Coyle, had done the requisite research to identify the actual gun parts from which the original blaster is made. The bulk of the gun comes from the receiver and bolt of a Steyr-Mannlicher .222 target rifle. Inside that (and holding the actual bullets for the prop) is a Bulldog 5-shot revolver. This information was a godsend for *BR* freaks like me, except for the fact that these guns are not cheap. Together they cost me well

Top: Barrel, receiver , and slide of my third and final Blade Runner *blaster, midfabrication. Bottom: Blaster version 2 (middle), soon to be dethroned by blaster version 3 (top left).*

Accurate in every possible way to the original, except for the serial number. Most people replace the serial number of their Steyr with the correct 5223 from the original, but I liked the idea that mine would be different in that respect, that it would be mine.

into the thousands. I was making good money, but I wasn't making Dick Wolf *Law & Order* money. I purchased one of each of these guns.

This presented a problem: both guns required heavy modification to turn them into the prop I sought. With only one of each at my disposal, I had to be ABSOLUTELY SURE about every single cut I made. There was no room for error. At all. Steinschneider and Coyle had intelligently made castings of their gun parts to modify the castings as they developed the correct procedure for making accurate blasters. With the benefit of their careful research, and with information gleaned from the original blaster, which surfaced a few years back, I embarked on the meticulous, painstaking, multistep process of gunsmithing and fitting it together to make my blaster.

This is the other kind of tolerance that can be utilized in the creative process. Not the one based on performance, like with the giant water pump, but the one based on limited material resources. Without any "mistake tolerance," there is only one reasonable and effective substitute that will get you where you want to go: time. Time is the only kind of mistake tolerance you can build into a project when you have limited supplies or knowledge. You just have to go slowly. When it came to smithing this third (and final) blaster out of the .222 target rifle and revolver, I took A TON of time to get it right. In fact, it eventually took over four years to complete. It was a costly trade-off, but a necessary one. One that, in my younger years with less experience, I may have ignored at my peril.

Now, with the experience of many, many wrong turns righted and mistakes corrected, going slowly is my main method for doing something that I don't know how to do, in a situation where I don't have a lot to work with—tools, supplies, margin for error, whatever. I simply go slowly. Very slowly. Even slower than you're thinking right now.

A skilled craftsperson has amassed enough knowledge about their particular discipline to move at a speed that is cost-effective: they don't spend so much time on a job that they can't make any money doing it. But for those of us who are generalists or ambitious amateurs making things for ourselves, we can often substitute that knowledge with time. This is the great secret sauce for tackling the unfamiliar.

DESTINATION UNKNOWN

It's common for people to describe any sequence of related events they experience as something of a "journey." This implies that a trip is being taken along a road or path, with a starting point and destination. It's an apt analogy because movement is movement, whether it's interior or external. On an actual road trip, you most often know what the destination is, and the path you take to get there is a combination of guidance, guessing, memory, and, even with GPS units in every one of our pockets, some trial and error.

But what about when you're unsure of the destination? What does the path look like then? Usually, it has lots of tributaries, countless tributaries. Every one might be the right turn or it might be a wrong turn. How do you know which is which? You can assess whether the turn you've taken brings you more in line with the direction you know you're supposed to be heading. You can lean on experience. The more time you've logged on this particular "road," the more mistakes you've been party to, the more ways you've seen things go sour and head in the wrong direction, the more quickly you're able to get things turned around and headed toward what you believe is your final destination.

In an interview with David Sylvester, painter Francis Bacon said about painting, "One has an intention, but what really happens comes about in working." A painter as amazing as Bacon is saying clearly that his painting is a hunt in which he's never sure what he might catch. As a maker of any kind, with any project, you will never *really* know what your destination is. You know your starting point, you know roughly what your "problem to solve" is, and you can try having a whiteboard session about final goals to help you figure out what you'd like your destination to be, or at least what the rough outline of it should be. You might even figure out what you'd like it to "feel" like. It's good to have a North Star to move toward, in that regard. But it won't change the fact that nothing can quite prepare you for what it's like to set out along the path of creation only to realize that you are not going to end up where you planned. No amount of ideation, whiteboarding, storyboarding, gaming out the options, will show you your true destination, you will only know it when you arrive.

There is a very simple reason for this, one articulated in a different context by the nineteenth-century Prussian commander, Helmuth von Moltke, when he said, "No plan survives first contact with implementation." Put another way: How many of your projects turned out EXACTLY like you intended? How many went as smoothly as you expected? Mistake free. Distraction free. You had all the time in the world. In my experience, the answer is pretty close to none. And I would argue, as makers, that this is exactly how we like it and why we do it. We pursue the things we do precisely because we can't know the outcomes ahead of time. If we knew exactly how it would go, why would we proceed? What would be the point?

Kurt Vonnegut was fond of saying, "Travel plans gone astray are dancing lessons from God." I think that's the actual mystery solved of the creative process. That's what keeps us coming back for more after every bonehead error, every wrong turn, every misfire, every disproven assumption, and, ultimately, every finished project. We just have to loosen our tolerance with ourselves and give ourselves the space to make all those mistakes. It's the only way we'll learn, the only way we'll grow, the only way we'll make anything truly great.

SCREW > GLUE

Jamie has always thought that making things involves taking large chunks of stuff and making them smaller in precise ways. That is true, but it's only half of the equation. It ignores assembly. Very little of what I've built over the years is monolithic—just a single chunk. Most of the time, I build things in components, then attach those pieces together as I go. So yes, the component parts are pieces that have been made small in precise ways from larger chunks of material, but eventually they will be assembled to create much larger and more complex objects than any of the raw source materials.

Assembly is an engineer's way of saying "put things together." This process of joining things together is always fraught with hard-to-see hazards, particularly since it usually happens last, and many of the most precarious and exacting operations within the physical making of things happen really close to the very end of a build. This makes the cost of failure high. If you messed up and ruin a part at an end-stage assembly, you have to do a lot of backtracking to get back to where you just were right before your misstep.

This is a common hazard for every maker, one I've been entrenched in hundreds of times over the years. And all the times I've had to backtrack, to build something again from scratch, have given me ample time to reflect on how to avoid these hazards in the first place. I've learned that a craftsperson isn't someone who doesn't make any mistakes. Master craftspeople are confronted with all the same problems and pitfalls as any other maker, they just have the experience (hard won via hardship—nature's primary learning tool) to see these hazards coming from farther away than the newbie. And in seeing farther, they have more time to get out of the way. A big part of that wisdom is knowing how best to assemble things.

When you need to attach two things to each other, you can do it in one of two methods: mechanically or nonmechanically. Mechanical connections are things like screws and nails, nuts and bolts, rivets or pins, zippers or Velcro. It's any fastener that you can remove and replace in some kind of easy(ish) repeatable fashion that doesn't damage the parts they are connecting. Nonmechanical connections are much easier to wrap your head around: they're adhesives like glue (or tape).* Of the two, glue is very often the faster solution. Mechanical connections require more foresight and engineering to get them to work, and they definitely require more labor. But they also leave many more options open, because they're reversible, which is why I love all aspects of mechanical fasteners.

Nuts, bolts, threaded inserts, thread repair coils, cotter pins, keyways, dovetails, rivets, even hook and loop. Welded studs, leather rivets, grommets, lacing eyelets, 80/20 extruded bar. Mechanical fasteners make everything about an initial build take longer, and they require planning and foresight to make sure

* These delineations aren't exactly industry-wide designations, they're more *my* own working methodologies that rely on experience and common sense.

whichever fasteners you choose register to each other, and to the parts you're working with. But after they've been built in, mechanical fasteners make everything after that easier. They allow for disassembly, reconfiguration, as well as replacement. That trade-off, of spending time up front to save it on the back end, is one of those choices that becomes clearer with the more experience you get.

When Luc Besson's incredible film, *The Fifth Element*, came out in 1997, I became obsessed with replicating the ZF-1, the gun handled by Gary Oldman's iconic villain, Zorg. The ZF-1 shoots bullets, arrows, flames, ice, rockets, and even a net!—all within a small egg-shaped package. Its comical complexity drew me in like a moth to a flame. I went online hunting for any reference material I could find, in order to replicate it, and ended up meeting another builder equally obsessed, Shawn Morgan, at the Replica Props Forum (where I would end up posting the Broom's Box drawings from *Hellboy* a few year later).

Shawn and I worked together on a ZF-1 replica for years. Our goal was to make a replica entirely out of aluminum and resin, just like the original. We began by making tons of 3-D drawings to try and gather all of our understanding in a single place. It quickly became apparent that there would be at least 175 separate, custom little parts and pieces that needed to be water cut, laser cut, cast in resin, or hand-machined in steel or aluminum if we were to make an exact, complete model.

The fact is, I could have easily just made each piece look right and then glued them all in place, but early on I knew that being able to dismantle a build of this complexity was paramount to the final product looking and feeling exactly right. So we spent a number of extra weeks in the early stages, both wrapping our heads around the difficulty and also designing and envisioning how to attach everything together so it could be taken apart again.

The aluminum skeleton of the ZF-1, prepped for the myriad fasteners and mechanical solutions that would allow me to improve it in the years to come.

When it was all said and done, I think I used nearly every type of fastener I know about in this thing: there are water-cut aluminum frames held together with tiny machine bolts because, with so many things involved, space was at a premium. I cast and molded the egg-shaped shell with large screw bosses in them so that I could easily remove them to have access to the inside. There are rivets, pins, pressure fits, and threaded inserts. I can't report for certain all the different fasteners I tried or ended up going with on the ZF-1, but what I can tell you is that there's no glue holding it together whatsoever.

It's not easy to build this way. Many times a part was completed only to realize that the holes that had been water cut into the internal aluminum frame were the wrong spacing, and so a new set of water-cut aluminum parts was needed. I am pretty sure

I built this gun about ten times before I completed the hero version to my satisfaction.

I completed my ZF-1 in 2015, and it's STILL not done. To this very day, I'm adding details and functionality to it, and the fact that I can take it completely apart and put it back together again, with a whole bevy of mechanical fasteners, is what makes this possible. I don't regret a bit of the many, many iterations of each part I went through to make it both accurate and also modular. It's light-years better than having glued it all together.

PHILOSOPHY OF GLUE

As much as I prefer a mechanical solution, there are moments when only glue will do, but not just any glue. The *right* glue. Picking the perfect glue for a project is as critical as it is difficult. The glue aisles of a hardware store can be a mystifying cornucopia of material science and chemistry: five-minute epoxies; cyanoacrylate *and* accelerator; white glue; wood glue; all-purpose glue (which is NEVER all-purpose); paper glue; contact cement; weld bond; silicone.

The conundrum of choosing the correct glue can be an anxiety-inducing ordeal for a neophyte maker. It's a weird thing to think about when you consider that, as a kid, glue seemed both straightforward and magical. Wile E. Coyote spreads it on the ground to catch Road Runner and, boom, he instantly gets stuck himself! That's how powerful glue is. To children, glue is a mystical substance that simply "joins" things. Many adults carry this same mind-set into their work, right up until the moment that they ruin something they worked really hard to make and don't know how to fix.

To my mind, there are three cornerstones to the philosophy of glue: 1) glue is about joining things together; 2) glue is most often wet coming out of the package, and can't do its work until it's dry; and 3) not all glues are created equal.

If you're joining two things that are similar, your job is pretty easy. For wood, use wood glue. For paper, use a PVA like Elmer's glue. For silicone rubber, use a silicone-based glue. And on and on. In each case, what you are looking for is a glue whose physical properties, when set (hardness, flexibility, response to temperature, etc.), most closely match the physical properties of the materials you're joining. It's all about temperature, mechanical properties, and use cases. For most common materials, there's a glue out there that has been specially formulated to solve the problem.

Picture bricks living out their life span of a few decades on a chimney. The life of a brick is a stationary one. They don't move, so they don't require a flexible glue. The mechanical load on bricks is compressive. They sit on top of each other, not moving, and the overall weight is huge, so you want a glue with a high compressive strength. Imagine if you used a silicone glue to do your bricklaying: it might work for a row or two, but because its compressive strength is low, and its flexibility is high, your building would sag, lean, and eventually fall apart.*

And then there's the weather to consider. As temperature changes, so do the properties of materials: cold temperatures make things brittle and hard, and heat makes them soft and flexible. Temperature also affects the size of things: hot things expand in size, and for the most part, cold things contract. Heat makes things expand more than you think. The main cables that hold up the Golden Gate Bridge are nearly seventeen feet longer in the middle of a sunny day than they are in the middle of the night. That size change is different for all materials.

* Now that I've described it like that, I kinda want to see it happen . . .

Thus, bricklaying grout has been specially formulated to behave just like bricks when it's set. It matches the bricks precisely in its compressive strength, as well as in how much it expands and contracts during shifts in temperature.

Now let's join two pieces of leather. Unlike bricks, leather is highly flexible. And the use case for leather often includes lots of torsion and movement. Grout would be impossible to use for leather. It would literally be hard as a rock, and as the leather moved around, it would easily tear and peel itself away from the grout to which it was bonded. Fortunately for leatherworkers everywhere, we have contact cement. Contact cement is a rubber- and solvent-based glue, whose material properties so closely match leather that two pieces of leather properly glued together with contact cement are more likely to tear within the leather than where the two pieces have been joined.

This might sound obvious, to use a glue whose properties match the materials being joined, but it's not to many people, so don't feel bad if it's not obvious to you. Plus, there are bigger fish to fry, specifically figuring out what to use when you're dealing with *dissimilar* materials, like joining two things that are NOT alike in their properties, like wood to glass or leather to rubber. That's when which glue to use becomes a matter of experience, experimentation, and often expletives, because you'll need an adhesive that balances between the mechanical, and material properties of both components.

Let's take a tough example: glass and metal.

Picture gluing a piece of glass to metal, and then leaving it outside for a year. Over that year the temperature might shift by fifty to eighty degrees or more. Sitting there exposed to the elements, the glass and metal expand and contract with the shifting temperatures. In and of itself, this isn't an issue. The problem is that *they don't expand and contract identically to each other.* So if the glass expands just a little, while the metal expands a lot, the two surfaces will start to work against each other. An adhesive for vastly dissimilar materials needs to be able to move just enough to hold on to both sides of the equation, without being compromised by either.

For instance, when installing glass in office buildings, the construction industry uses industrial adhesives that never fully set hard as a rock. They always stay just a bit soft (the set consistency is sort of like dried chewing gum) and as such, hold on to both sides with equal strength. They need to, as the cost of failure of a falling piece of glass from an office building is catastrophic. The windshield of your car uses a similar type of adhesive.

While the consequences are rarely so dire for the typical maker in their workshop, the situation they face is the same. If you don't get the right glue, you risk structural failure and you risk ruining both parts being joined, sending you back more steps in the fabrication process than you'd care to consider.

THE GLUES IN MY SHOP

Everyone has glues that they prefer, and I'm no exception. Over the years I've learned about and incorporated many glues into my arsenal of adhesive solutions, and I'm excited to share them with you. Just please take into account that my knowledge of these glues and their functional properties is just that—functional. They are the conclusions of an experimental generalist. I am not a chemist, nor a physicist, nor a material scientist. And there are exceptions to every single rule I outline below. You may meet someone who tells you there's a better solution to every use case I outline. Go ahead and take their advice seriously and try it out for yourself. You may justifiably disagree with some of the glues for which I have great affection. That's fine! It simply betrays, in the best way, the limitless land of possibilities when it comes to the developed ingenuity of humans to modify their world.

AIR-DRY GLUES. This is the biggest category—glues that, when exposed to air, dry out to create a bond. These can be water-based, like PVA glues (Elmer's, basic white glues, wood glues) or solvent-based, like many "all-purpose" glues.

PVA glues are awesome and incredibly useful. Wood glue is one of the few things that does what it says it does just as well as you imagine it might. It produces a solid bond, with high strength and flexibility. White glues, like Elmer's, are also phenomenal when using them on the porous materials for which they're best suited—stuff like lightweight paper and corrugated cardboard. A new, thinner formulation of PVA glues (a main brand is called Mod Podge) has become invaluable for people who make things—props, costumes, models—out of foam. These glues are water-based, which means they have no noxious chemicals that stink up your house, and are easy to clean up.

ALL-PURPOSE GLUES. These are a constant source of disappointment to me. I've been let down so many times by "all-purpose glues" that they deserve only the briefest mention here. Sometimes they're plastic solvent-based, like Duco Cement, sometimes they're silicone-based, but they're almost always in my opinion a temporary solution. Some people swear by them, I'm not one of them. Your results may vary.

CONTACT CEMENTS. I really like contact cements. These rubber-based glues also flash off their solvent base to dry, but are applied in a different way. Contact cements are mostly used by applying the glue to both surfaces being joined, letting the glue set for a short while, but not too long, and THEN joining the surfaces together. Blow-dryers are good for accelerating the drying process with contact cement. I use them all the time in my one-day builds for Tested.com when whatever I'm building involves a fair amount of glue-up.

When properly applied, this stuff can work miracles. They put shoes together with it. That should tell you something. Indeed, this family of glue makes tough, strong, yet flexible bonds suitable for everything from shoes to foam to attaching posters on board. For porous materials, I often use two coats of the glue on each side. In the right application, this can be one of the most powerful glues for gluing dissimilar materials together as well. Contact cement comes in tubes, cans, and even spray cans. I've used them all. My favorite contact cement is called Barge glue. A favorite among leatherworkers, I've found it to be properly tenacious. However, the cheap generic contact cement you get from your local mom-and-pop hardware store has rarely failed me, either.

HOT GLUE. Hot glue is a thermoplastic substance. That means it is highly reactive to heat and cold and moves a lot with temperature variation. At room temperature it is a moderately flexible plastic that comes in stick form. When you feed the glue stick into a glue gun with a heating element, it comes out as a thick, hot, honey-like liquid. As it cools, it sets.

For quick and dirty builds, hot glue can't be beat. For anything that you want to last, I would tend to avoid hot glue like the plague. (I've had things that were hot-glued together just fall apart while mounted on a wall, for instance.) Hot glue works best for porous things like wood and cardboard—it's AMAZING for cardboard—and terrible for nonporous things like metals and glass.

Hot glues most often come in a clear or translucent form, but can also be bought in colors. I've used red hot glue for doing fake wax seals on theater props like old-timey envelopes. There are also low-heat variants of hot glue, which can be great when you're gluing something like Styrofoam or foamcore board that melts easily. Hot glue can also be used to make castings. I know of a theater show that needed chicken legs for a dinner scene and the scenic department used hot glue squirted into a silicone mold made from an actual chicken leg. While I'm certain it didn't *taste* just like chicken, the end result weirdly felt just like chicken.

EPOXY 2-PART GLUES. These are thermoset glues wherein you use two separate liquids—a resin and a hardener—and mix them together to create an exothermic (heat-generating) reaction that chemically sets the mix. Two-part epoxies are often brittle glues, but there are more flexible formulations out there as well. They often come in the classic "5-minute" glue form in a pair of tubes with a plunger. Unlike a thermoplastic glue, epoxies are permanently set once they've hardened. They cannot be remelted with the application of more heat.

Epoxies are a great nonsmelly product to use with fiberglass. Boats around the world are built using glass matte and epoxy resin. All the ships in the original *Star Wars* were put together largely with epoxies. The big downside of epoxies is that they can be quite bad for the human body, so make sure you use gloves and work in a well-ventilated area when your project calls for it. And if you're using large amounts of epoxies, for coating something or fiberglassing, use a chemical respirator. JB Weld is a particularly fine tube-based example of these glues. There are even (plausibly apocryphal) stories about it holding a motorcycle crankcase together long enough for the bike's rider to get to a repair shop.

EPOXY PUTTY. Also a thermoset adhesive, epoxy putty is a big family of glues that can join things but can also be used as a maker material in its own right. There are versions for plumbers that can be applied around leaky pipes. Others are made to repair leaks on boats. There are versions formulated specifically for metals, for woods, and for plastics. There are super lightweight versions, too. Because they come in a claylike form, they can also be ideal for one-off constructions.

Like the glues, epoxy putty has two parts, often in two different colors, each the consistency of clay. Take the two parts, knead them until you have created a third color and can't see either of the original colors, and then put the putty to use. Some set fast, others set slow. I've used epoxy putty to make dollhouse bathtubs, fantasy gun grips, and even file handles. Once they're set they can be sanded, cut, and even screwed together with common woodworking tools.

CYANOACRYLATE (CA). Often referred to by the brand name Krazy Glue, this class of glue, cyanoacrylate, is the soul of the special effects industry. It was originally formulated as emergency battlefield sutures during the Vietnam War and I know model makers who swear by CA glue for stitch-worthy cuts (I've never tried it). Lorne Peterson, one of the original model makers on *Star Wars* and an old friend, was the one who discovered CA glue as an Eastman Kodak product and introduced it to the ILM model shop. It's import in the special effects industry cannot be overstated.

CA glue is highly versatile and comes in liquid form with varying degrees of viscosity, from the super-thin (very tricky to use right) to the super-thick gap-filling kind, to the newest version, flexible CA glue, which I'm just starting to use, and like a lot. All of them can set when exposed to air or be accelerated by adding a "kicker." Once set, it becomes a hard acrylic that tends to be brittle, so you want to be careful.

Super-thin CA glue deserves some extra attention here because it's so thin (like vodka) that it flashes quickly and sets almost instantaneously, way faster than any other CA glue formulation. It's amazing for things like ceramics, where it can wick into the porous ceramic and leave almost no trace of itself. But if you have your skin anywhere near that crack, it'll wick underneath your finger and glue you to the very thing you're trying to fix. The fact is, few glues can get you into more trouble more quickly than a super-thin CA glue. I have glued hero props to my own hand with super-thin glue more times than I care to admit.

It's not just your own skin you need to worry about either. You have to be vigilant with the entire surface of the object to which you are applying it. Working on a commercial for Jamie back in the mid-90s, I spent a week making brass corners and filigree for an incredible glossy lacquered box that my colleague Lauren built out of hardwood. On the day of the shoot, while trying to hold down a recalcitrant brass corner, I used super-thin CA glue and it wicked into the joint, and then ran right down the front of the prop, on the camera side. I still remember the feeling in the pit of my stomach. Jamie was pissed, and when Jamie gets pissed, he doesn't show it much in his voice or his manner, instead his head turns bright red. He's like a human thermometer, with a mood head. He ended up having to use crayon wax that he mixed and formulated to match the lacquer finish to hide my glue streak from the camera, and he had to redress it for every shot. His head was bright red the whole time!

Long story short: super-thin CA glue, beware.

CA accelerators are what is referred to as "kickers." They are solvents that can be added to typical CA glues to accelerate their setting time from minutes to mere seconds. They come in spritz bottles and spray cans, and can also be applied with a needle applicator. They can be amazing when you're doing quick and dirty model work, especially in films and commercials; you just have to remember that solvent-based CA kickers can often have a deleterious effect on paint jobs and on clear plastics (beware of kicker and polycarbonate!). So as with any unfamiliar process, do a test first on scrap material to understand a CA accelerator's effect on your build. You don't want to do anything you can't fix later.

Little known fact: baking soda is also an excellent accelerator for CA glues. It kicks them almost instantaneously, and it doesn't really smell at all. Sprinkling a little bit on glue you've laid down also makes it immensely stronger. I've used baking soda and CA glues to create gusset-like welds on the inside of styrene boxes that made them incredibly strong. I have known plenty of model makers over the years who can't abide the smell of solvent kicker and only use baking soda.

WELD-BOND GLUES. These are a special class of adhesive. A weld-bonded glue melts both sides of the bonding equation and then dries out effectively making one part. This is why it's called a "weld." Weld bonds are absolutely fantastic for gluing together acrylics and other plastics. Model airplane glue is a thickened type of plastic weld bond, but it also comes in a much wetter form used for making things like acrylic boxes. The glue that plumbers use for joining PVC piping together is a weld bond. I love weld-bond glues, they create strong joints, and they do their work fast.

There are different formulations of weld bond for different types of plastic. There are weld-bond glues for ABS, polycarbonate, and PVC. I tend to use styrene and acrylic mostly in my shop, so my glue of choice is Weld-on 3. For styrene scratch building or putting small pieces of plastic sheet together, there's nothing better than a bit of Weld-on 3 and a brush.

That's it. Those are 95 percent of the glues that I use in my work. There are exceptions to every rule, like I said, and there are some really wild and wacky variants out there that are worth looking into depending on what you make, but knowing these glues and what they do is only part of the equation.

THE PRIMACY OF SURFACE PREP

Once you've chosen your glue, and you have the materials you plan to join, it is never going to be simply a matter of applying the glue and sticking the parts together—not by a long shot. If the parts you're gluing aren't clean, they won't adhere. If they don't have sufficient grip, they may slide or peel right off each other. To ensure a good, solid bond between materials, you'll have to do what's called surface prep. That is, you'll need to prepare each surface to be glued.

This part of the process is as, or more, important than the glue or the materials themselves. In every case, you'll need to clean the surfaces to be joined of dust, oil, and moisture. If they're painted, you'll want to remove some of the paint where the glue will go (especially if you're using a weld bond). Sometimes, you may even need to abrade the surfaces to increase adherence, as is often the case with shiny mirrorlike surfaces.

Try it. Glue two shiny mirrored surfaces together and after the glue is set you'll likely be able to peel the two pieces apart. Why? Because the shiny surface has the lowest possible surface area, that's why it's shiny. If you look at it in a microscope you'll see the same thing: a flat, glassy surface. But if you looked at that same mirror surface after it's been abraded by sanding, you'd see microscopic hills and valleys. These hills and valleys provide vastly more surface area for the glue to get into and get some more grip on the material. This is what's called giving the glue some "tooth." Without those hills and valleys, without that tooth, the glue might just slide off the surface. That's why sanding smooth and nonporous surfaces is so crucial to creating a powerful bond.* I've even used a blade to deeply score two surfaces to be joined, to give real deep tooth to the bond.

OPTIONS, OPTIONS, OPTIONS

Ultimately, glue is more often than not a temporary measure, a stopgap, or a compromise. In its capacity as such, it gamely fulfills the truism of compromise: it leaves both sides of the equation slightly wanting. The problem is that there is no perfect solution for every material bonding situation, only compromises. This isn't to say that glue isn't amazing when it works properly, because it is. I just don't like that much of the time it's hard-to-impossible to reverse a glue joint. I am allergic to that kind of one-way operation, in all areas of my life really.

I live in San Francisco, a city famous for its terrible traffic. But one of the things I love about this city is that no matter where you are, no matter how terrible traffic is, most of the time you have options. And keeping my options open is one of the reasons that during high-traffic times I like to avoid the freeways. Freeways limit options. City streets, while often taking longer to drive through, leave options open at every juncture. Glue does not come with options. Glue something and when it's set, pulling it cleanly

* When sanding for surface prep, use a heavier sandpaper, like 100-grit or even rougher.

apart again does not work. And trying to make it an option is often catastrophic.

That's the reason I prefer mechanical solutions. They can be undone. Whatever I'm putting together can be pulled apart again without damaging the construction. Like navigating the hilly labyrinthine streets of San Francisco, it takes more engineering, more fiddling, and definitely more time. But the trade-off is more options. And I want options. That's the space I like to exist in as a maker.

9

SHARE

Sharing what I know is a personal mission. It's a key part of how I balance the scales for the incredible gifts I've been given. Whatever success I've enjoyed in my life has always been directly related to those who've supported me, and to all of the amazing people I've been lucky enough to meet, know, collaborate with, and learn from. As a maker and storyteller, I see myself as part of a continuum, going back to the beginning of humans using tools and telling stories, and continuing forward into infinite possible futures. Sharing information is the fuel for the engine of progress.

However, I've run into many people who don't believe in sharing their work, or more specifically, that sharing their work, their methodologies, their custom processes, even their enthusiasm, incurs a direct *cost* to them. I have been tangling with this kind of cynicism and parochial, scarcity mind-set for most of my career. I understand it intellectually, but emotionally I will never understand it. Why would you not want to share the things you love? Why would you not want to share the cool things you've made? Or the triumph of a challenging project that you've overcome with your friends? Why would you hide the knowledge you've acquired over the years or pretend that your hopes and dreams aren't worth shouting from the rooftops?

In my experience, the more you give away, the richer you will be (to paraphrase Paul McCartney).

I've felt this way as long as I can remember. My bias has been toward openness: open source; freemium pricing; open door policies; you name it, if it involves getting more creative knowledge and more tools into the hands of more people, I have been all for it. And like most great things in my life, this, too, began with *Star Wars.*

SHARE WHAT YOU'RE INTERESTED IN

I was ten years old when *Star Wars* came out, right in the sweet spot for the world it created to have an indelible impact on me. I watched it for the first of what would be many, many times from the backseat of my parents' Toyota Corolla at a drive-in theater in Cape Cod. These were far from the best viewing circumstances—all I could see of Darth Vader and Obi-Wan's duel was the lightsabers because the combatants themselves were blocked by the front seat headrests—but still, my memories of that night are clear. I remember my dad hating it. "Well that was such a piece of CRAP," he said as we pulled out of the drive-in and headed home. He considered it obvious, simplistic, and boring. I was astonished by the fact that we did not share the same opinion of the movie we'd just seen, and then concluded in about six femtoseconds that my father, for the first time in my short life, was clearly wrong on this one. I keenly remember the shot of shiny gold C-3PO in the very beginning of the film, standing next to a shiny silver C-3PO, and thinking to myself that one shiny robot was amazing, but TWO!?! I also remember having a sense that I was seeing the tip of the iceberg—that there was a much bigger universe behind this five-thousand-foot strip of celluloid that went further and deeper than I could see, but I was immediately prepared to follow it wherever it went.

For my eleventh birthday the following summer it was all about *Star Wars* toys. Action figures, blasters, and, of course, lightsabers. I remember opening the figures I got so fast that I lost Leah's blaster. I'm still kinda sad about that. And then I added to my *Star Wars* play the other gateway drug for making (besides cardboard): LEGOs, but augmented with paper towel tubes (elevator shafts, natch). I even made my own Death Star from LEGOs, replete with secret doors, evil and good droids, and a smaller planet that it could destroy.

As the *Star Wars* frenzy continued to build and began to take the shape of the entertainment juggernaut it would eventually become, stories from behind the scenes found their way into the pages of the only periodicals I really cared about: *Cinefantastique*, *Famous Monsters of Filmland*, and *Fangoria*. I devoured every single article I could find about the *Star Wars* universe, and luckily there were a lot. I pored over close-up full-color photo spreads of Wookiees and Jawas. Most formatively of all, I learned that there were *people* whose job it was to build the ships and props that made this incredible movie so engrossing. I would have worked that out on my own eventually, I'm sure, but at the time I just couldn't believe that there was employment to be had and money to be made in doing the kind of thing I'd already been doing in my uncle's workshop, and my father's studio, and my room since I was five years old.

When you're a kid, learning this kind of information produces a perception-altering, paradigm-shifting epiphany. The things you see out in the world, on television or the big screen, for the most part they just . . . *are*. And the assumption is that they always have *been*. You don't question them. Droids are droids, Wookiees are just Wookiees. It's one of those comfortable tautologies of innocence. The idea that there is a man inside that 'bot, or underneath all that fur, who is performing and getting paid for his effort and going home to his furless human family at the end of each day, who are not robots, is like breaking the fourth wall on childhood. It registers like an earthquake at the center of your worldview, which unleashes a tsunami of realizations. Foremost among them, at least for me, was HOLY HELL, IS THAT SOMETHING I WANT TO DO! Right then and there, my career goals shifted from LEGO designer to "person who makes stuff on *Star Wars*"—a goal that I am amazed, to this day, I was eventually able to achieve.

The truly great service that magazines like *Fangoria* and *Cinefantastique* provided was to offer their writers and readers a platform for learning about and sharing their passions. As a kid who struggled to make friends and who spent most of his time playing by himself in his room, I wasn't just surprised and excited that there was a job in this stuff, I was comforted by the fact that there were more people out there who might share an interest in the things that I loved. Maybe none of them were in my class, or even in my school, but they were out there somewhere, which turned them into a beacon that I could aim myself toward. Wherever those people were, that was where I wanted to go.

Turns out, I didn't have to go that far at first, because a lot of them were just down the Saw Mill River Parkway in Manhattan. At Tisch and at the film school, the ability to openly share our mutual interests in things like science fiction and space, movies and model making, tinkering and complex problem solving, was the foundation of a large creative community on which friendships were built and where opportunities were found. I got my first crack at making film sets and props on David Bourla's senior thesis film, for instance, specifically because we spent all those hours at the movies, and in his living room, over coffee and cigarettes, talking about and obsessing about our shared love of movies, sci-fi, and fantasy.

That process of opening yourself up and sharing the cultural commonalities you care about can be a pathway to learning more than you could ever imagine. Admittedly, it can make you vulnerable, too, because the people in whom you confide this interest can either laugh at you, as happened to me growing up, or, worse, they can accept your interest and simply not care, as happened to me in my early twenties when I took several stabs at finding work in the effects industry.

The effects houses where I landed in New York City were not great work environments. I found they were often exploitative, sour, divisive places. They expected a production assistant to work fourteen hours per day (or longer!) for fifty bucks flat. Even in 1986 that was pretty terrible pay. In my mind I understood that paying my dues was an exchange for learning new things. However, what I got instead was a crash course in all the ways someone can tell you to stay in your lane and mind your own damned business. No one in those shops seemed interested in sharing ANYTHING about what they did or what they one day hoped to do, probably because the hope had been drubbed out of them by the people who came before them—people just like them. It was dispiriting, but I persisted in this openness because I knew it would pay off. It could equally be possible that I was an annoying kid (likely) who pissed off my supervisors with his off-topic eagerness (plausible), but the end result was that my early forays into the world of special effects didn't feel that special.

I hit pay dirt when I met Jamie several years later in San Francisco. He was a terrific boss in so many respects: he paid well and also gave reasonable raises, he provided good feedback, and most importantly, he registered my curiosity and let me use his shop to learn about pretty much anything I was interested in. A dizzying array of processes are available in a standard effects shop, which needs to be a kind of Swiss army knife of a shop: lathe work, milling machines, mold making, metal forming, pottery and casting supplies and tools, foam carving, animatronics, vacuum forming, pneumatics, painting. These were just a few of the skills I saw being used on a regular basis in the shop. I told Jamie what my goals were and what I was interested in. I asked him if I could experiment in the shop in my free time, and I asked him and his partner, Mitch Romanauski, for help whenever I was out over my skis. They repeatedly said YES and thanks in great part to them I now possess a bevy of weapons in my creative arsenal.

If you are struggling to figure out how to move forward in your job or your environment, whether you're in a creative field or not, my best advice is to figure out any aspect(s) you find interesting, and share that interest with colleagues and bosses in an effort to learn more about it. I have a shop assistant named Mel who works for me in the summers. At one point Mel told me that they wanted to learn more about painting and weathering. It just so happened that I had recently picked up a 3-D printed kit of a storm-trooper rifle on Etsy, so I asked Mel to put it together and prime it black. I then spent an hour or so walking Mel through three different specific techniques for applying a realistic metallic look to plastic, and how to think through how such a gun might age over time—how the edges get worn, how to see that the weathering looks better when it is uneven (a particularly hard thing to learn, because things don't weather in uniform ways). Mel dove in and did a terrific job with it, taking my instruction and applying it to their own unique skill base. Mel turned out to be a natural. This immediately increased the projects I could give Mel. My teaching moment provided a direct benefit to both of us, something I'd learned long before from Jamie. A good boss will encourage this type of chutzpah and support a culture that produces more of it. It's good business sense. As an employer myself now, I appreciate when the people who work with me want to learn more, and express that, and I'm happy to provide the space for them to increase their skill base. A shop is a possibility engine. More skills among its inhabitants increase their effectiveness and boost the efficiency of the engine as a whole. A net benefit for everyone.

SHARE WHAT YOU'VE DONE

Much of my success has depended on my ability to do quality work in a timely manner. For the opportunity to do that work, I have relied upon other people hearing about it first. I frequently benefited from word of mouth. In the maker world, word of mouth is everything. EVERYTHING. And it begins with yourself, specifically, talking about the things you've gotten really good at and sharing the evidence.

After a few years in the San Francisco theater scene eating humble pie and building my skill base, I naturally drifted toward a specialty: props. I became known for being good at designing and building props as well as solving complicated mechanical prop and set problems. Eventually, I was hired by the prop department for the Berkeley Repertory Theatre to work on the first stage production adapted from Maxine Hong Kingston's amazing book, *The Woman Warrior.* There were ghosts in the show that caused chairs and plants to move by themselves. The theater at the Berkeley Rep wasn't big, it fit four hundred people at most, and

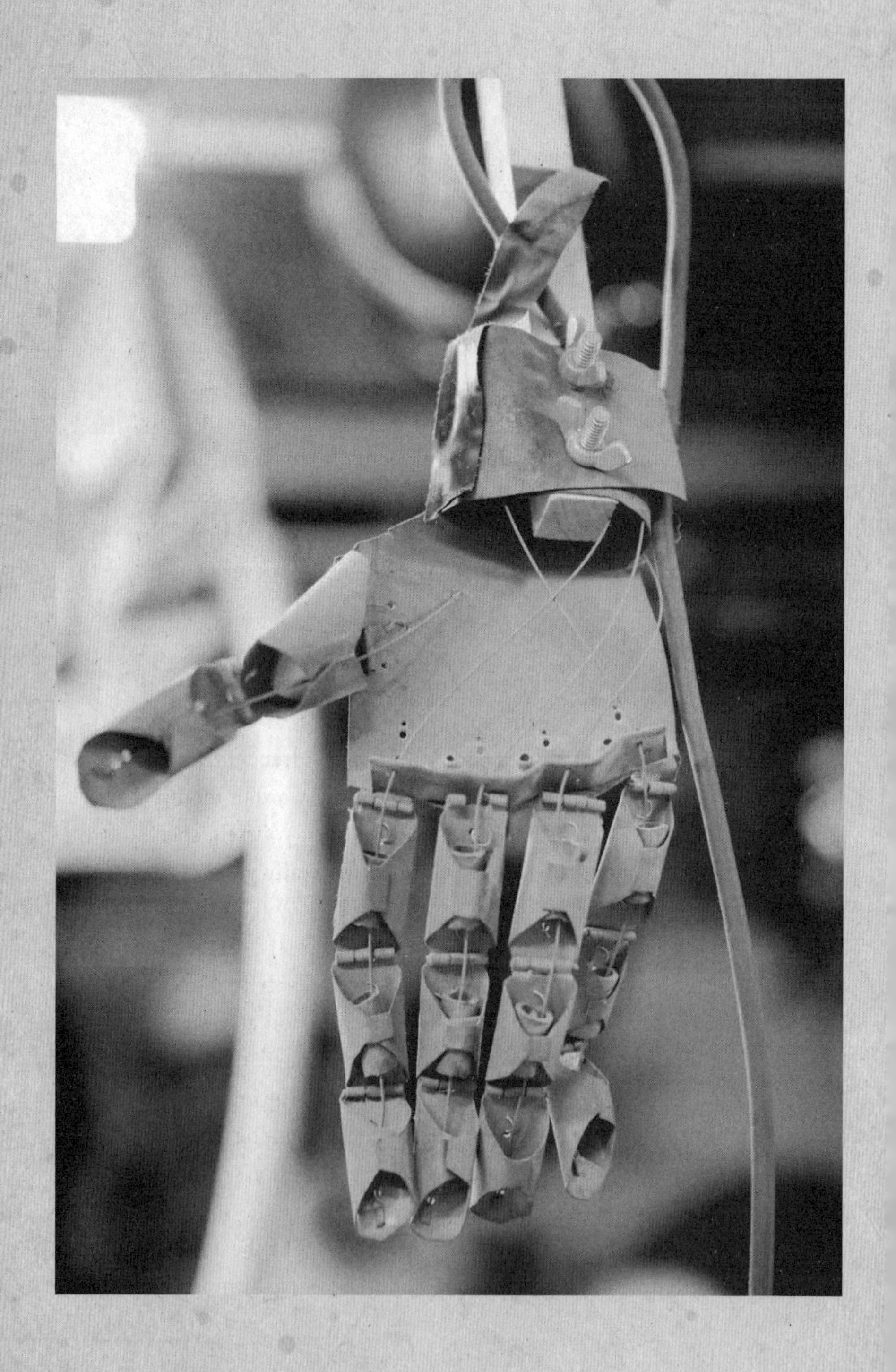

the stage jutted out into the audience so that there was no hiding from them. Performance, costume, makeup, set dressing, props—it was all totally exposed. There was no faking it.

My primary task on that production was to make a robotic easy chair and robotic palm tree that could fool everyone, from front row center to the back of the house. After dozens of hours of thinking and sketching and stopping and starting, I ended up making a pair of compact, high torque, robust, remote-controlled robotic chassis. They worked like a charm and were among my proudest accomplishments during my time in theater.

It wasn't long after the Berkeley Rep job that I got a call to interview with Jamie at his shop. Typically in creative fields like ours, standard interview practice was to bring in a portfolio filled with high-resolution, full-color images of your work. The interviewer spends a few minutes leafing through the pages, those minutes expanding into torturous hours in your mind, until finally the interviewer closes the portfolio and asks you a bunch of questions that give no indication whether they liked your stuff or not.

Opposite: Made from paper, fishing line, and metal linkages, this mechanical hand gave me a leg up on my interview with Jamie. Above: You can't tell by looking, but these robotic motors are massively heavy. They operated the easy chair and the palm tree from The Woman Warrior.

It's not a great system. So instead, I committed to showing up to this interview with a suitcase full of the actual things I'd made. It's one thing to share glossy, two-dimensional representations of what you've made. I figured it's something else entirely to put that stuff right into the hands of the person who decides whether or not you get the job. As a supervisor myself now, I know that a ton of useful information can be learned about a person from seeing, in person, even just a single thing they've made.

Pictures rarely communicate all the necessary information about how skillfully a job was completed. Being able to feel the object in their hands and to walk around it and inspect it from all angles at once, gives a prospective employer the fullest of possible pictures. Plus, at least for me anyway, having the objects in front of me spikes my own enthusiasm for them. Coupled with my ability to speak to every aspect of their physical character as well as the projects that produced them, I am able to tell a much better story about myself and my work than an employer can intuit from a book of photographs. My enthusiasm overrides my nervousness about being interviewed.

When I met with Jamie, I could only bring pictures of the mechanisms I built for *The Woman Warrior* because the robot chassis were massive and too big to carry. But I also brought a suitcase full of objects I'd built, mechanical hands, engineered linkages, painted models. We spent about an hour combing through the pile—he getting a sense for who I was as a maker, me telling the story of how they came to be. And the rest, as they say, is history.

The fact is when you interview for a job you rarely get to showcase all your skills. That's true regardless of the job you seek or the field you work in. Sharing what you've done—*showing your work*—is the closest you can get, because each of those objects is an embodiment of the skills you've acquired and the lessons you've learned over time. It can be a mobile app or a five-thousand-word article about why Whole Foods is full of angry people, or the mechanical assembly for a robot tree. It doesn't matter.

"You will eventually use everything you've ever learned," someone once told Steve Martin about his budding comedy career.* Indeed, every job I've ever had, from unpaid programmer for public access cable television, to library page, to busboy, to graphic designer, to actor, to scenic painter, to toy designer, has come into play in what I do professionally.

Of course, this instinct to share what you've done extends beyond finding a job. It applies equally to starting your own business, finding collaborators, or just planting your flag in the ground as a maker. Sharing your work announces your presence. Being vocal about your achievements is an investment in yourself. It doesn't have to be big and showy. Don't be a blowhard (I say this for my own benefit as well). You don't need to be perfectly polished. Start a blog or an Instagram account. Go to 'cons and meetups and exhibitions. Give yourself a name. Embrace the noun (Maker, Painter, Writer, Designer) by sharing with the world evidence that you've been living the verb (making, painting, writing, designing).

Just don't be a bore, and definitely don't believe your own bullshit. Believe me everyone can tell the difference between someone who just talks the talk and someone who can walk the walk. Before I calmed down in my late twenties, I could monopolize a conversation with the best of them. I regularly projected my enthusiasm on to my audience, which is to say I was sure they would be as excited about my projects as I was. It produced some uncomfortable silences. I'm not proud of those naive, self-involved days. Being a better listener rather than a

* From *Born Standing Up*, Martin's incredible memoir about his life as a stand-up comic.

"wait-to-talker" is something I'm *still* working on, but it doesn't change the fact that listening to people share things about which they're sincerely enthusiastic can be both thrilling and inspiring. You never know, that person to whom you're showing your first pair of homemade gloves might someday remember that glove making is something you do and seek you out to make more some day in the future. I got lots of work over the years with just this kind of approach. So go ahead, share your work.

SHARE THE CREDIT

What I didn't know when I walked into my interview with Jamie all those years ago was that my name had preceded me. People in the theater scene whom I had worked for and with were talking about me and my work to people like Jamie. When they talked about their successful productions and proudest achievements, they did not take all the glory, they gave credit where they believed it was due, and I was one of the beneficiaries of that largesse.

I recognize my fortune in this regard. Having spent nearly twenty years in television, I assure you that credit hogging in Hollywood is a contact sport, and as an industry it is hardly unique. One of the world's most popular social networking sites, Reddit, is jam-packed with people posting original content, and just as quickly jammed with other people reposting that content as their own. It can be nerve-wracking to never be sure if you're going to get the credit you deserve for the hard work you've done. But that doesn't mean you should stop giving it where it's earned. The fact remains that to make something special, to create anything great, it really does take a village. Nobody does anything new truly by themselves. As social beings, we interact. As explorers, we push ourselves and each other. As problem solvers, we learn from those around us every bit as much as we learn from our own successes and failures. No amount of credit hogging is going to change those facts. To locate your success entirely in yourself is to ignore, and therefore disparage, the contributions of all those people who helped you get to where you are.

I take pride in my work, and I have no problem saying, I MADE THAT, but I also believe strongly in publicly sharing that pride with all those who were part of the journey. It gets the word out about what we've accomplished, what my collaborators have done as individual makers, and what is possible when you put individual egos aside and come together for a common creative purpose. That's how things like my space suits, like the NASA ACES suit, or my pair of *2001* Clavius suits get built. It really does take a village.

This is the ethos that Richard Taylor, Peter Jackson, and Jamie Selkirk were trying to cultivate when they named their New Zealand–based effects company, Weta Workshop. Richard's first effects shop was called RT Effects (as in, Richard Taylor Effects). Peter Jackson is a massively famous director. Jamie is an Oscar-winning editor. They easily could have named the business after themselves, negotiating over the arrangement of the initials or the order of the names in the title, as is often the case with law firms, architecture firms, and other collective enterprises. But Weta isn't about them. It's not even about the native prehistoric cricket-like bug from which the company took its name. "We call ourselves Weta, so everyone feels like they're working under a collective yet singular banner," Richard told me on a visit to his shop in Wellington in early 2018.

Weta has eleven different departments. Unlike most shops in the United States, the makers at Weta regularly transition between departments. "It's a co-op of people working together who enjoy not specializing, who enjoy being entirely collaborative," Richard said, "to the extent that we will create sculptures that pass from artist to artist to artist, so that over time the collective is adding something richer to it than any one individual."

This is a beautiful philosophy and I love it, not least because it accounts for ego in a healthy way. "We want our people to be egotistical, we want them to take pride in what they do," Richard explained, "but it can't be to the point that they're excluding others from doing the work with them." Because if their sculpture work is any indication, the whole at Weta is always greater than the sum of its parts, and the credit for that belongs as much to the company's owners as it does to its artists, which is exactly how it should be.

We make a lot of stuff in my little shop in the Mission District. We also collaborate, commission, and trade skills, talents, and techniques with many other makers around the world. There's a Buddhist sentiment, one of Buddha's five reminders, paraphrased by Thich Nhat Hanh that speaks to the heart of what motivates me in our collaborations: "My actions are my only true belongings: I cannot escape their consequences. My actions are the ground on which I stand." In my time in the theater scene, at ILM and Colossal Pictures, on *MythBusters* and in my shop, I have stood on the shoulders of great artists, fantastic builders, and truly ingenious makers. To not acknowledge their contribution to my success, to not give them their due, isn't just malpractice, it's an injustice. It is the kind of act for which the karmic consequences are, indeed, inescapable.

SHARE YOUR KNOWLEDGE

When I got started in Jamie's shop in the '90s, I did a bit of machining and worked alongside a talented machinist/engineer named

Chris Rand who didn't say much but did his job exceedingly well. I was incredibly green when I met Chris and it was clear he didn't think much of me as a machinist, nor should he have, but he was still very helpful in his way.

On any given day I'd set up a rig on the old Bridgeport milling machine, thinking that I'd clamped everything down pat, but to make sure, I'd steal a glance over at Chris. I knew he wouldn't *say* anything to me about my setup, but I also knew he was watching me carefully, and his opinion would make itself known in his facial expression or in a subtle gesture. If he didn't like what I'd done, he'd just barely shake his head—Chris-speak for "You are clearly about to destroy something." So I'd take my rig apart and set it up again. And again. Each time I'd get the headshake. Finally, after maybe three or four attempts, I'd get the compliment I was looking for: a slight shrug. This was a genuine encomium from Chris. It meant I hadn't totally screwed the rig up. It was his way of saying, "Well, at least you're not a complete idiot."

I could probably count on two hands the number of words we exchanged related to work process, but in all those shrugs and headshakes, Chris shared a mountain of machining knowledge with me. Isaac Newton once said, "If I have seen further it is by standing on the shoulders of giants." What he was talking about was the foundation of all human progress: the sharing of knowledge. I've had many mentors over the years, Chris included. I've learned countless things from each of them. The one thing they all had in common was an understanding that generosity with one's knowledge is paramount for our species, because knowledge is power, and the most powerful thing you can do with knowledge is give it away. My mentors were not cynical about up-and-comers like the guys in the Manhattan effects shops. They did not possess scarcity mind-sets that would inevitably shrink their worlds into nothingness. They were open and they benefited mightily from it. In fact, many of them are still working today on some of the greatest film and TV franchises the world will ever know.

And yet, I have encountered more than my fair share of people who just aren't interested in sharing what they know, who are not interested in progress in our field. In the early 2000s, I worked next to a wonderful model maker who spent weeks implementing a specific plastering process, to great effect. I asked if I could take some pictures and document the many steps he was going through to achieve the finish he was getting (I'm circumspect about the details to protect his identity) and he said, "Sure! But I'm not going to tell you everything . . ." He felt that this particular skill was key to his employability, and thus if I knew it, I'd be his competition. He admitted that his reticence was "weird" but that this is what he believed, so he was going to remain mum about the important bits. While he is certainly not alone in this fear, we'll just have to agree to disagree.

When I completed the Mecha Glove build, I presented Guillermo del Toro's glove to him at Comic-Con in 2014, but something inside me was saying that I still had some unfinished business to attend to. I felt I needed to help others achieve the same pleasurable outcome I'd enjoyed. So I made a road map to the process I used to create it.

During this period, I was often flying to different colleges around the country to do appearances with Jamie. Over the course of a couple dozen long plane rides, with the help of Photoshop and my brain still overflowing with the thousands of details of the project, I turned the lists and the drawings I'd used to make the Mecha Glove into actual, usable art. The finished product is a poster of every small piece, bit, and bauble that comprises this incredible prop.

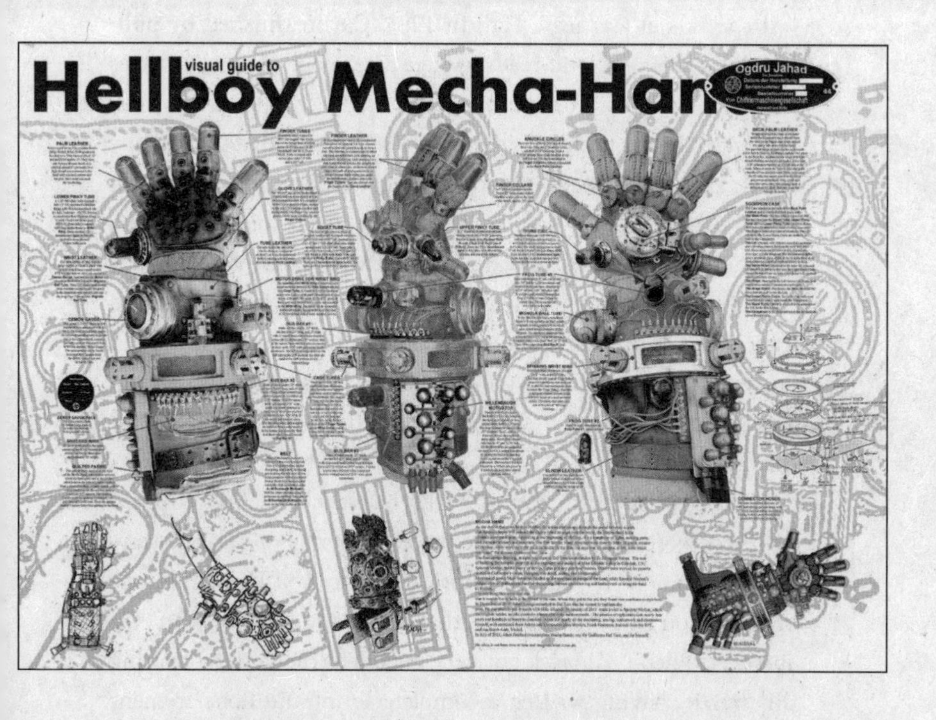

I did this because I like my lists and renderings to have a life after I've used them, and because I like to share. The idea that someone could use this info to satisfy their own desire to obtain this object pleases me to no end.

I've been told that mine is not the normal perspective on stuff like the Mecha Glove project, but it turns out that I've always been like this. Back when I first compiled all of the dirty words that George Carlin said you can't say on television, I had this sense that I was doing more than satisfying my own curiosity. I had put together an important document, one that should be made more accessible than an HBO special. So I carefully wrote out each curse word on a separate index card. I added definitions where I could. I added a few of my own that I knew. In one of his last specials at Carnegie Hall in 1983, Carlin finished by pulling out a legal pad and reading page after page of all the curses that he'd ever heard or could think of as the credits rolled. They numbered in the hundreds if memory serves. I added them to my collection, sorted them into alphabetical order, and put them in a tiny metal file-card box. Why? I'd love to say for safekeeping, but that's not where my head was as a twelve-year-old. This was the first full "collection" that I'd ever compiled, and I wanted to leave it in a condition that other people could utilize. For what? I had no idea. But that wasn't my concern. This was genius wisdom and it needed to be shared. That perspective would guide me for the next forty years.

I discovered a fun commonality between how I approached the dirty words, and one of the pieces in Stanley Kubrick's traveling exhibition. If you're a Kubrick fanatic, you know that one of the famous unfinished films that he tried to make was an epic about Napoleon. Kubrick got very far down the path to making this movie, having written a complete script and done tremendous amounts of costume research and location scouting. But there was one part of his research I didn't know about.

In the exhibit is an unassuming card catalog. It was compiled by the history department of the local university in England where he was working on his Napoleon project. In the drawers of this card catalog, fully cross-referenced, is compiled every single moment of Napoleon's life that has been chronicled somewhere—every person that he met, every place that he went, everything he did. The very audacity to imagine that such a database was possible, let alone achievable, just blows me away. And the fact that the Kubrick family would share it with the world thrills me every time I think about it.

SHARE YOUR VISION

If you ever want to create something great, you will have to collaborate with other makers. You will have to get good at sharing not just your ideas, but your vision for those ideas. What they look like, how they will be used, why they need to be made. Basically, what is your dream scenario for this thing that you want your collaborators to help you build? You have to be able to share what's in your head with them, and be able to receive what they've got in their head when they hear your idea.

When I made Heywood Floyd's lunch box from *2001: A Space Odyssey,* I shared my vision for it with Tom Sachs. He is just as obsessive about Stanley Kubrick as I am, and we agreed it would be a really fun object to dive into together. So at the same time that I was building my own version from scratch, Tom was as well,

Tom's lunch box and thermos is on the left, my lunch box is on the right. We each built two, and sent each other the second box, so we each have a complete set—a shared vision with entirely different outcomes. Welcome to creativity.

except not as a perfect rendition of the filming prop like I was. Tom was looking to replicate the prop in his own inimitable style, out of plywood with hardware-store details. We had a common idea with a shared vision that differed only at the margins, margins that defined our individual aesthetic as makers. The results were awesome.

A couple years before Tom and I went down the *2001* rabbit hole, I visited Guillermo del Toro on the set of *Pacific Rim*. I was gobsmacked by the sheer scale of the endeavor he'd undertaken. He was in charge of hundreds of people who were all charged with working together toward a gigantic shared vision of huge robots fighting massive monsters—world building on a nearly inconceivable scale. How does one even begin to captain such a gigantic ship? How is it possible to manage a group of dozens of artists to keep to a cohesive vision? At dinner that night I asked Guillermo how he did it.

"You have to give everyone complete autonomy within a narrow bandwidth," he replied. What he meant was that after you get their buy-in on the larger vision, you need to strictly define their roles in the fulfillment of that vision, and then you need to set them free to do their thing. You want the people helping you to be energized and involved; you want them contributing *their* creativity, not just following *your* orders. Giving them creative autonomy rewards their individual genius while keeping them oriented to the North Star of your larger shared vision.

Whether you are the captain of your creative ship or the lowest swabbie relegated to the poop deck of the project, the fact remains that none of us is an island. We are each of us part of a community, never more so than when we are makers, creating new worlds out of our imaginations. It's nice to think we can do it alone. It pushes all our ego buttons to consider ourselves the singular genius. But experience shows all makers that every success is a shared success, and every shared success is an investment into the culture that produced the success in the first place. I believe the world is a better place when we're all pulling on the same rope.

As a maker, it's up to you to decide what you're going to do with all the knowledge you accumulate. Are you going to hide it? Are you going to pretend that you alighted upon all your insights by divine providence? Or are you going to share what you have learned? Are you going to open yourself up to the people in your orbit and show them who you are, what you love, what you've made, what you know, who's helped you, and what you plan to do with all this to make the world a better place to live?

I know what my answer would be. What's yours?

SEE EVERYTHING, REACH EVERYTHING

A shop is not simply a place to make things. Yes, it's where we collect our materials, our tools, our notes, and our half-completed ideas, but it's also a manifestation of how we think about organization, project management, and working priorities. Packed with our personal histories, the shop is where we get to enjoy the illusion that the universe has some order and that we as creators can pretend we have some measure of control over things. A shop is a meta-level tool for telling our stories. It is an autobiography of our whole experience as makers. It's where the problems we choose to solve have stakes big enough to challenge us. It's where our successes and failures play out in microcosm. It's where we encounter the world, and confront our own minds.

Every shop, in this sense, is an individual philosophical discourse about how to work, one held up by personal beliefs that, like anything, evolve with time and experience and wisdom, but are always a reflection of you. They are reflected in the answers to questions we can always be asking ourselves as makers: What kind of work do I do? How do I like to work? What tools and materials do I use most often? Do I like it calm or crazy? Do I like shelves or bins or pegboards or drawers or racks or all of the above? The job of a maker is to learn the answers to these questions and to understand how they affect the shape and philosophy of the shop where they are brought to bear, so that the maker's own evolution is always forward and never stuck in place, or spinning in a circle, or trying to adapt themselves to the philosophies of others.

Nick Offerman, the famously mustachioed star of the NBC sitcom *Parks and Recreation*, has been a maker his entire life. He grew up about an hour southwest of Chicago, in a huge farming family whose members were all makers in their own right. "Farmers have to be incredible mechanics and biologists and animal husbandry experts and carpenters," Nick recalled when we talked late one summer morning about his shop.

Nick's first shop was probably the farmhouse he grew up in as a boy. But even that is stretching it. "With great regularity, my dad and I would be in charge of the firewood for the three wood-burning stoves in the farmhouse. That meant a chain saw, an ax, a sledgehammer, splitting wedges," Nick said. His first workshop, in reality, was the forest and his philosophy was raw efficiency.

It wasn't until he moved to Chicago in his twenties to work in theater as a scenery builder that he was able to put together what one might reasonably call a shop, out of unused warehouse space that his friend's landlord had been keeping unoccupied as some kind of tax dodge. "He was just sitting on it and I said, 'Oh, well, you know, there's a lot of malfeasance in Chicago. I would happily live there and build scenery in it and I could be twenty-four-hour security for you.' He was *also* a bullshitter and he smiled and said, 'I think we can work something out.' "

You can imagine what the shop of a young scenery builder and bullshitter-par-excellence might look like in an illegal live-work warehouse space. "I had a throwaway table saw from a big theater, a chop saw—those were pretty much the only stationary items you would have for a theater shop—other than a full set of hand tools, Dewalt's first sexy bumblebee six-pack of battery-powered tools—trim saw, cordless drills, Sawzall, cordless jigsaw—then, a router and sanding equipment. That was pretty much all that was required. When you became deluxe, you would get a little compressor and a brad nailer and a stapler."

When you're young, your first shop is likely to be similarly rudimentary, yet armored by the arrogance of youth. It will reflect the internal wrestling match between the three parts of your developing psyche. If you're anything like me, your id will bully your ego into a tag team and they will come off the top rope onto the back of your superego, driving its face into the mat and rendering it effectively unconscious (no pun intended), incapable of regulating expected and accepted behavior. For Nick, that meant dismissing something as elemental as a woodworking bench to hold all his scenery-making materials. "Out of ignorance, I looked down on woodworking itself because I was a scenery guy," Nick admitted. "Building scenery, the name of the game was speed. Something that held your material while you worked on it seemed way too precious to me." Instead, Nick would just grab a couple of saw horses and some clamps and let it rip. This spoke directly to the heart of my process as a young maker. "That required a great deal more strength and agility, though," Nick continued, "and ten percent of the time something would fail and I would fall through the wall I had just built." This spoke to me, too.

It often seems like our first work spaces are an energized mess of chaos and creation that we will insist at the time we have our arms around, but with some distance—both physical and temporal—we will realize was retarding our creative output in one way or another.

Indeed, my first shop was not even a shop at all. It was a studio apartment in Park Slope, Brooklyn, that the landlord let me live in rent-free as long as I moved out when he decided to sell the building.* The inside of my apartment looked like a topographical map made from literal DUNES of junk. Like my mind at the time, the studio was bursting at the seams with jumbled mounds of raw materials, half-formed ideas, and reclaimed garbage. Quite literally, I garbage picked constantly. I fancied myself a found-object artist in those days, and New York City in the mid-1980s was a garbage picker's paradise. New Yorkers threw out the most amazing stuff, from pachinko machines, to motorized dentist chairs, to nineteenth-century steamer trunks. It was intoxicating.

I lived in that Park Slope studio for over two years, until an incident with an escaped boa constrictor and a snake-phobic neighbor brought my time in Brooklyn to a close. I made a bunch of sculptures and other cool things, all from the contents of the mounds I'd formed from my scavenging exploits. I spent hours every day on my knees hunched over a collection of parts on the floor that I'd clawed out of the garbage dunes, intent on turning them into . . . something. The first sculpture I ever sold I made there. I would be lying if I said those weren't a blissful few years, despite my abysmal work habits, but it would also be a lie if I didn't admit that by engaging in the time-honored practice of moving back home with my folks when I was twenty-one years old to gather my wits about me, I could finally see the limitations of this kind of process.

* This is an illegal practice called "warehousing," of which I was more than happy to be a beneficiary.

A typical session on the floor of my first "shop" in Park Slope, circa 1986.

Upon moving home, with freelance graphic design gigs looking spotty at best and the horror of Manhattan effects shops barely a distant memory, I decided to set myself up down in the basement. I'd arrived with all my tools and the "materials" I could fit into my car that I didn't think would get me immediately kicked out of the house. I grabbed whatever else I could find lying around the basement that my dad wasn't using and then I spent several days organizing all of it into what would become my first real honest-to-goodness big-boy workshop.

I organized the space along two axes. One, I needed to maximize the limited amount of space, which meant pushing everything against the walls, stacking stuff on shelves and under tables, and hanging things off pegs and hooks and nails. Two, I wanted to

Above: from my Water of Faces series. Left: Home Run *was my first sale, to Lee Lorenz, the* New Yorker *cartoon art editor, cartoonist, and a family friend.*

Inspired by the amazing collages David Hockney was making at the time, this was my attempt to work in that form and capture the look and feel of my shop (1989).

tailor the space to my needs both for construction and for inspiration. This was an ongoing process, because at twenty-one years old, I really had no idea about what I required for either, and I had no insight into what method of organization would work best.

Not all organizational methodologies are created equal. One could be spotlessly organized, with everything put away and labeled and color coded, and it could feel like a prison with the walls closing in around you. Another could be equally organized but a bit more open and exposed, and it could untap creative genius like no other space you've worked in. My goal was to find a balance. When I got close, not only did my productivity and my

This is what I thought a real artist looked like. So hipster! (1989)

inspiration increase, but I enjoyed a renewed sense that I was a real artist. I wasn't a Brooklyn wannabe. These weren't flights of fancy I was indulging, like when I was a kid; these were worthy, artistic endeavors.

I spent about a year back at home with my parents. It was tough, yet it was also a very productive time: I made *a lot* of sculptures. Then, in the spring of 1990, when a good friend in San Francisco invited me to come be his roommate ("I don't relish being roommates with you but you couldn't possibly be worse than my current roommate," was how he put it if memory serves), I moved across the country and never left. My first shop on the West Coast was a single bench in the living room of an apartment that we shared in the Western Addition in San Francisco. Over the next few years, I worked in and visited dozens of shops, each of them giving me ideas and inspiration, and informing how I would eventually build my own. I was picturing one shop to rule

I know it looks tiny, even for San Francisco, but this was a great shop. The pod from The Fifth Element *ZF-1 gun is visible on the left, obscuring the bottom right corner of the Coca-Cola sign.*

them all, when in reality, there would be many, of many different sizes and persuasions.

The smallest shop I ever put together was truly tiny: only eight feet wide by twelve feet long, but I packed a lot of use into that space. I did most of the early gunsmithing work on my *Blade Runner* blaster there, as well as much of the machine work on the ZF-1 from *The Fifth Element.* The largest I've built is my current shop (aka The Cave) in the Mission District, at around 2,500 square feet. I've been there since 2011, and its organizational scheme has never stopped evolving.

I can see now with distance the similarities between all the shops I've built over time. They are tied together, in part, because I've been doing roughly the same kind of creative work for many years, so the tools, materials, and organizational strategies have

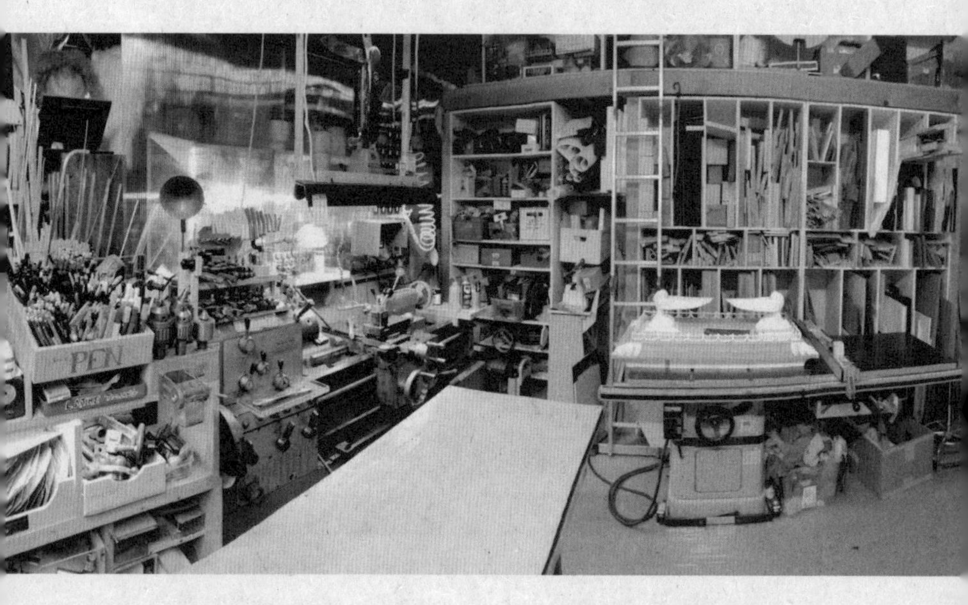

stayed roughly similar as well. What truly unifies my shops, especially as I got more experienced, is that they are each built on two, simple philosophical pillars: 1) I want to be able to see everything easily; and 2) I want to be able to reach everything easily. When I talk about a shop being a discourse on how to work, this is what that looks like for me.

SEEING EVERYTHING EASILY: THE VISUAL CACOPHONY

If you had asked me back in my Brooklyn days what my ideal work space was, I would have described a huge open loft with a workbench and a couple dozen of those big rolling canvas factory carts that were used for mail or laundry back in ye olden times. Each one would be full of a different kind of material: one full of motors, one for LEGOs, one for electronics, one for plastic toys with cool shapes, and on and on. Basically, my ideal work environment would have

The Cave, circa 2014.

traded the chaos of stationary mounds of stuff for mounds of like items in buckets with wheels. It was the fitting encapsulation of my working process at the time: haphazard and organic.

As the Sleepy Hollow shop took shape, I realized that I actually liked the organic part of my process. I wasn't interested in seeing all my tools and materials stacked up neat and tidy, I liked a more natural orientation. I find comfort in spaces that are difficult to take in at one glance, but I also need to be able to make sense of the space. The mounds of junk were not allowing my mind to synthesize all the information my eyes were taking in so that my hands knew instinctively where to go and what to grab.

I have a term for this aspect of my organizational sensibility: visual cacophony. I think of it a bit like listening to a symphony orchestra warming up. Music isn't being made, per se, but there is a mellifluous energy that is recognizable to the ear, because each piece and each section is working through the same scales, the same piece of music in preparation for a performance. Once the conductor comes out and leads the orchestra through the sheet music—each piece on time, in proper proportion, in the right order—that's when the music is made.

It is the same for me, as a maker, with the visual arrangement of my shop. In the Park Slope studio, my dunes of detritus were like the sections of an orchestra holding one sharp note as loudly and for as long as possible. It was deafening. At Sleepy Hollow, the organizational logic of my shop slowly began to speak to how I liked to work—fast and stream of consciousness—at which point it turned into an aesthetic goal for each shop thereafter.

It is my growth as a maker that drives how that philosophical pillar is articulated in my shops and how the aesthetic goal is achieved. With every new skill I acquire, I also accumulate different tools, different glues, different paints, and materials necessary to practice that skill. As my skill base increases, so, too, does the contents of my shop and the options at my disposal for constructing each project. Don't misunderstand, I always have a clear vision for everything I want to make and a solid plan for building it, but there is more than one way to skin a cat; there are a thousand roads to Mecca; and there are countless ways I can cut, join, paint, finish, or assemble the pieces of a project. As someone who enjoys working fast, plowing ahead with ideas as fast as I can grab rags, plaster bandages, drill bits, and spritz bottles and paints to implement them, the more important it is to have all those tools and materials visible, out and in front of me, at all times.

This is also why I have a love-hate relationship with drawers. Let me tell you my philosophy about drawers: F*CK DRAWERS! Drawers are where stuff goes to die. Drawers lure you into a false sense of security by "helping" you put things away and making the shop look "cleaner." But really, by putting stuff out of sight, they remove it from your visual field. They flatten the visual cacophony into a monotone, or worse they mute it altogether. When you put something in a drawer, how sure are you that you'll remember this is the drawer that houses the stuff you put into it? How clearly did you label it? Is it grouped with similar materials? Out of sight, in my experience with drawers, truly does mean out of mind, and in a shop where I want to be able to find everything at a moment's notice, that is bad.

On the other hand, one of the things I love most dearly is to take a drawer that looks like this:

These are all my blade-holding devices and their concomitant blades.

and using foamcore, a snap knife, some hot glue, and about an hour of my time, turn it into this:

This picture fills me with calm every time I look at it.

I have created customized foamcore inserts like this for many of the drawers in my shop. I'm not done. Not sure I ever will be. It's a slow process because first I have to accept that the drawer is necessary, then I have to commit to *what* goes in the drawer before I memorialize the contents' placement with some level of permanence. Drawers organized like this bypass enough of my hatred for them to keep me using them because they allow me to quickly and easily identify things like that one weird tool I might use once a month but is the ONLY tool that can do the job for which it was selected. They also help me take stock so that I don't run out of things at the most inconvenient time possible. Most of all, custom inserts make it easier to see everything that should be in a drawer, but isn't, and needs to be tracked down, which cuts down on the biggest problem of all: feeling like you keep losing something and then going to the store to buy it again, and again, and again. It also helps me to commit to one of each type of tool I need in active circulation. As a lifelong maker, I've gathered, in some cases, many iterations of the same tools. I might have three different combination squares in a bin labeled "Rulers," but putting it into a drawer means I commit to the best one. And actually I often give the others away or place them in traveling tool kits, which makes the entire shop more efficient.

Drill bits, brushes, markers drawer.

Not everyone would agree with my bias toward visual chaos, of course. At ILM, one of the common practices at the beginning of every job was to grab a free four-by-eight-foot table and cover it with brown butcher paper that you secured with black tape. This gave you a neat, clean work surface, which they believed engendered neat, clean work habits. That principle was imbued throughout the ILM model shop, and I took it with me when I left . . . it just stopped at my drawers.

Don't get me wrong, there is a place for drawers in a shop. Drawers are good for tool and material types that tend toward the small, the specialized, or the diverse: blades, drill bits, markers, pencils, rivets, washers, Allen wrenches, glasses, brushes. They're just not to be taken for granted or used too liberally, because if you're anything like me, you don't want to just store stuff, you eventually want to retrieve and use it as well.

Allen wrench drawer.

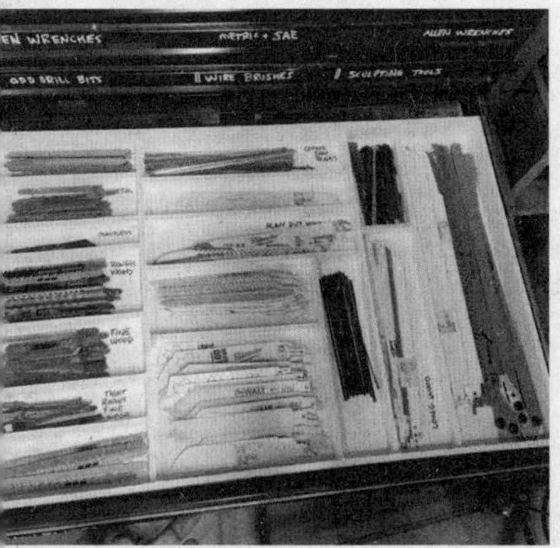

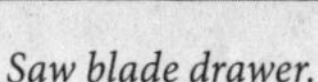
Saw blade drawer.

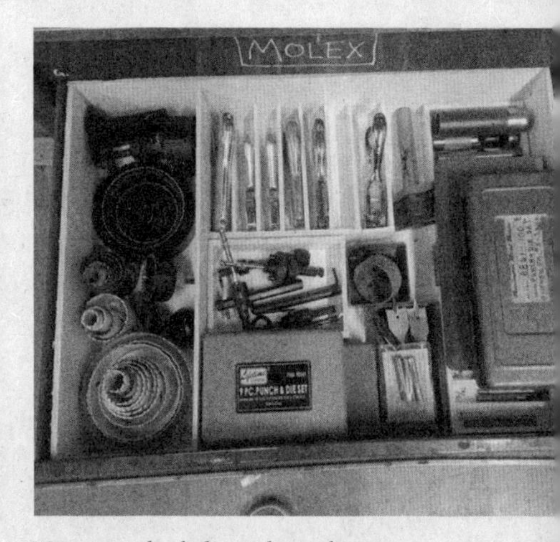

Ways to cut big holes in things drawer.

To do that efficiently with drawers requires a system, and I tried to play nice at first. When I moved into The Cave, I went on Craigslist and found a five-foot-tall stack of classic crackle-finish Kennedy toolboxes. Kennedy is the machinist's toolbox. They're expensive, though deals can still be found, but truth be told that didn't much matter to me at the time because what I really liked was the idea of adding this status symbol to my shop.

I filled this behemoth's three stacked toolboxes and twenty some-odd drawers with all my smaller hand tools—dozens upon dozens of them. Every drawer got a label. A few got inserts. None of it helped. There were too many choices. The labels, even in the largest possible font, were hard to read. More to the point, I COULDN'T SEE ANY OF THEM.

I gave the Kennedy stack a good four years to prove itself, but eventually, as a natural enemy of visual cacophony, I had to get rid of it. In its place I built a rolling, five-runged, five-foot-tall ladder rack with twenty holes drilled into each rung big enough to

3"
1"
1¼"

accommodate the handles of every tool I'd kept in the Kennedy stack. I literally turned the storage drawers inside out. Each tool has a place that's fairly obvious, they are all in sight, and I can get to any of them five times faster than it took me just to figure out which drawer they used to be in. I have virtually instant access to every kind of pliers, wrench, tweezer, grabber, nipper, and cutter you can imagine. It is functional, visual cacophony in its highest form. Also, remember how I said drawers are where stuff goes to die? I found so many tools I thought were lost in the shift from that Kennedy stack to the rack I built. The process of removing that particular stack of drawers confirmed all my feelings about drawers.

REACHING EVERYTHING EASILY: FIRST-ORDER RETRIEVABILITY

Over the years, each one of my shops, big and small, has been a refinement on this idea of visual cacophony. Each has been carefully, incrementally adjusted to my working style, my pace, and the way I think through materials. My current shop is the highest-level iteration of this, and the ladder rack is its epitome. It not only services my desire to see everything easily, it also delivers on my need to be able to reach everything easily, wherever in the shop I might be working, without having to move anything else out of the way to achieve that access. I call this "first-order retrievability."

If visual cacophony is the philosophy of my shop space, of *where* I work, then first-order retrievability is the philosophy of my process, of *how* I work. This will be different for every maker, of course. Nick Offerman's shop doesn't prioritize instant access

The inaugural ladder rack. I have since built a half dozen more. I have too many tools.

to everything, it reflects his process as a furniture maker and his preference for large-format woodworking.

"The layout and flow of a shop always has to do with the specificity of what that shop fabricates," Nick said. "In a furniture shop like mine, which is sort of exclusively different types of furniture and production items, you have your sheet goods right by the roll-up door for easiest accessibility. Then, you just try to create a flow so that your material can pass across the joiner and right next to that is the planer. It easily turns the corner to then be ripped on the table saw. All the sanders and shaping implements are in one area. I'm lucky enough that I have two sixteen-hundred-square-foot bays. So, one of them is all the machines, where the magic happens, clustered together for optimal vacuum dust collection. Then the second room is the assembly area where all the clamps and big glue-up tables live for finishing. I'm very fortunate to have that much space. Usually, in many shops you have one table and that has to perform all those functions."

That's exactly right. Nick and I are blessed with the amount of space we've been able to carve out for ourselves. If you work in a typical shop, you have to figure out how that "one table" is going to "perform all those functions." How does it get organized? How do the functions get prioritized? There is no one right answer, but there is a wrong one: whatever doesn't work best for you, personally, as a maker.

It actually took me awhile to figure that part out for myself. It wasn't until the early 2000s, a couple years into working with Jamie on *MythBusters*, but a full quarter century into my own makerdom, that I realized how differently I work than others. In those early *MythBusters* days, we'd do a lot of building in his shop, and I found that I was getting exhausted by the whole process. Literally, exhausted. Like my legs were tired from walking, because I was walking miles every day inside Jamie's shop retrieving tools.

Jamie's shop was, and is, a possibility engine. Every process you can imagine is available within his five-thousand-square-foot space: vacuum forming, pottery, fiberglass, welding, casting, carpentry, rigging, painting, airbrushing, mechanical design, animatronics, robotics, you name it. Like every shop, his has its own organizational philosophy, and it's driven by how Jamie likes to work. There is a room for welding, a room for electronics, a space for carpentry, a spray booth for painting. And in each of those areas, he has tools germane to those disciplines. So in the woodshop, for instance, there's an entire cubby devoted to over two dozen different hammers. But then I noticed something: that is the ONLY place in the whole building where you can find a hammer. That might not seem like a problem in theory, but you know where else hammers get regular use? The machine shop, where I spent a lot of time during our *MythBusters* tenure. When I worked in Jamie's shop, every time I needed a hammer in the machine shop I had to walk all the way to the woodshop, to the cubby where hammers were, grab one, and walk back.

This might sound like a trivial issue to you, but for me it violated one of my core creative principles: the efficiency of first-order retrievability. I know that sounds somewhat hypocritical coming from someone with a shop that looks like a carnival exploded inside a movie set and then got flung against the walls like a Jackson Pollock painting. But how something looks is not the same as how something works, and for me, Jamie's layout didn't work because I was wasting time. And there's nothing I hate more when I'm working than burning time getting tools when I could be holding them in my hand and using them. There were often moments, late into the day, or deep into a build, where my frustration with Jamie's setup overwhelmed me, and instead of making yet another trek to the woodshop, every tool in my immediate vicinity became a hammer—to great effect, I might add.

In my shop, by contrast, not only do I have rolling ladder racks everywhere, but I have multiple sets of everything I use on a semi-regular basis. I have three complete sets of T-handled Allen wrenches, for instance: one for my lathe, one for the mill, and one out in the middle of the shop in the general tool area. I use them constantly for things like tightening and adjusting my setups. It's a small time savings in each individual operation, but over the aggregate I'm sure that having those three sets of Allen wrenches on racks right nearby has saved me probably ten to twelve hours of walking around every year that I've had them. To my mind, the "extra" sets have paid for themselves many times over.

In those early *MythBusters* years, almost always in fits of pique, I often found myself asking Jamie, through exhortations to the heavens, WHY WOULD ANYONE NOT WANT A SECOND SOURCE OF HAMMERS CLOSER TO WHERE THEY MIGHT BE USED? But then as I got to know Jamie, I realized that I was asking the wrong question. Like Nick Offerman, Jamie is a farm boy. He was raised on an apple orchard. He grew up riding horses and driving tractors. He has carried with him into his professional life a Midwest farmer's work ethic that I would describe, in the most positive way, as plodding and methodical. It wasn't that he might not "want" extra hammers in his shop, he simply didn't think he *needed* them. What's more, I am quite sure that walking all the way across the shop to get a hammer is useful to Jamie. I'm positive that he uses the time while he's walking to think carefully about his next step.

Once I realized that Jamie's shop spoke to a specific set of values that he grew up with, I realized that my shop was no different. The design and layout of The Cave values speed and iteration. It values my stream-of-consciousness method of working. The rolling ladder rack was a revelation in visual cacophony because it allowed me to turn all my drawers inside out so I could see everything. But really it was the final *evolution* of my philosophy on first-order retrievability, which I had been moving toward since I was an eighteen-year-old found-object artist in Brooklyn.

Back then, working on my friends' student films over at NYU, I noticed that I was frequently leaving my studio with a bunch of tools dumped into whatever toolbox I had lying around that fit the number of things I had to bring with me that day. Then, once on-site, I would spend half my time riffling through the toolbox to find whatever I needed for nearly every task. I'd be constantly riffling through the toolbox that was the size of a toaster oven, trying to find something as big as a hammer, and not succeeding until I dumped the whole thing out onto a table.

The mounds of junk in my studio were unwieldy, but at least they made visual sense to me, as each mound was its own type of material. These tools, though? They were just a source of constant frustration, until one day I was walking down the street in the East Village and I came across a used leather sample case—the kind a traveling salesman would carry. The unique thing about this sample case was that it was wide and deep, but also tall—tall enough, in fact, that most of the tools that I carried with me on a regular basis could stand upright in it. This was a mini revelation. An upright tool takes a fraction of the floor space that a horizontal tool does and is infinitely more accessible. I immediately went to the art store and bought some upright pen and pencil sorters. I adjusted their dividers until they could hold things like vise grips, pliers, and small handsaws. Then, after every gig, I would come home and reorganize the layout of the case based on what I had learned from my work that day and what I was going to need for the next day.

When I left for San Francisco, the sample case came with me. As a freelancer, the case and the tools it contained were my livelihood. When it eventually started to break down and I needed to find a suitable replacement, a friend was gracious enough to trade me his grandfather's old doctor's bag for a conga drum I had. The bag's high center meant that, just like the sample case, I could store many of my important tools upright on both sides of a sorting divider fixed down the midline. The longer tools could then lay down horizontally on either side.

My grandfather was a surgeon, and I've always loved medical equipment and antiques in equal measure. The idea that an object like a doctor's bag, with its wonderful worn-in leather look and feel and the story that patina tells, could also be useful for transporting my tools? I was in. Over the next few years, my

Opposite: It may seem ridiculous to devote a whole page to a picture of my tape storage, but, in truth, tape has vexed me for years. I spent twenty years figuring out how to best store rolls of tape, until the answer arrived in early 2017, when I was in a theater on a stage tour and found that the stagehands stored their dozens of rolls of tape on shelves. Flat! I did some quick math in my head. This was indeed the most efficient way to pack tape. One could visually see all of it and yet have access to any roll with minimum disturbance to the organization. It was salvation from the dark side of tape: With that big hole in the center, tape lures you in with promises of quick access; it begs to be threaded on a spindle. "Put me on a large pipe!" tape practically screams at you. But that way is madness, and if you submit then it will have you. Because if you need a roll of 1/2" pink tape that you've placed in the middle of the pipe, you're now pulling down pounds of tape, removing a bunch of rolls from the spindle to get the one you need, then putting it all back. It's not a useful solution; it's a shoulder workout. Ugh. When I saw how those stagehands stored their tape, I made a quick sketch so I wouldn't forget and the first thing I did after I came home from the stage tour was build this shelf. And, yes, I recognize that this is a ridiculous amount of tape.

doctor's bag underwent constant refinement, then replication. At a flea market in San Francisco's Bayview District, I came across a second identical doctor's bag and filled it in short order.

It was around this time that I got hired at ILM. I showed up on my first day with my doctor's bags in their fullest splendor. I made quite an impression. I became known as "that guy with the insane toolboxes" who works very, very fast. But not *quite* as fast as I could be, according to my first supervisor, Michael Lynch, who remarked that I was spending too much time bending to the floor to retrieve my tools. He suggested that I put them on scissor lifts. It was a stroke of pure genius. (I knew *Star Wars* wasn't the only reason I wanted to work there so badly!) I went home at the end of that day and spent all night making a pair of rolling scissor-lift platforms. In fact I built them twice that evening, as the first set I made collapsed immediately under the weight of the doctor's bags.

I loved these contraptions so much, but up to that point I had never considered the limit of what my leather doctor's bags could hold, or that I was shortly about to reach it. The combined weight of a couple hundred hand tools, plus a rolling scissor lift bolted to their bottom that restricted the leather's ability to distribute the weight, slowly caused the bags to begin to fail. Unfortunately, my love for their form factor and their deep history could not stop physics. It could not overcome the limits of the bags' mechanical design when faced with the demands I put upon them.

When one of the handles finally tore off on a Friday afternoon, I went into my little shop in the Mission District and spent the weekend rebuilding the doctor's bags from scratch out of aluminum plate and pop rivets. It took thirty hours and more than seven hundred rivets, but on Monday morning I showed up triumphantly to ILM feeling like the Moses of makers come down from the mountaintop with two aluminum boxes containing the Covenant of the Creative. Okay, maybe that's a little hyperbolic, but I definitely felt like a badass.

And this was only the first iteration of these toolboxes. I kept refining them. Every time I picked up a better technique or a new tip or, most importantly, a new tool from one of the artists at ILM, I added it to my skill base and, if appropriate, to my toolboxes. Thus, they were never truly "finished." Just as I currently do with my rolling ladder racks, and with The Cave itself, they are always a work in progress.

TAKE A STEP BACK TO MAKE A LEAP FORWARD

Surprisingly, my toolboxes reached a high level of development before I ever truly asked myself why I needed my tools laid out that way in the first place. I never stepped back and looked at them and thought to myself, *What is the optimal configuration of these tools given the space provided?* I was too busy working to do something that contemplative. Rather, I was simply following my nose, and adjusting my environment to fit my work habits. If something didn't work as well as it should have on Monday, I changed it for Tuesday. If Tuesday's configuration slowed down something else, I changed it yet again for Wednesday.

Every maker goes through that process with their shop if they're paying attention and trying to get better. It's an incremental process

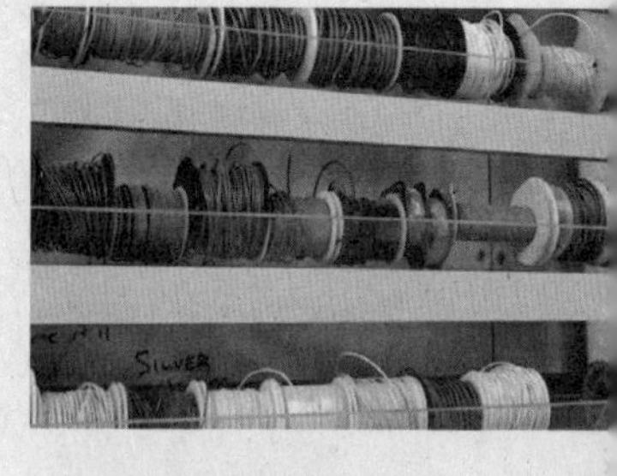

In service to both visibility and access, I got rid of spindles, aka The Devil's Rotisserie, and in addition to storing tape horizontally, I also placed wire on angled shelves with individual feed-through slots.

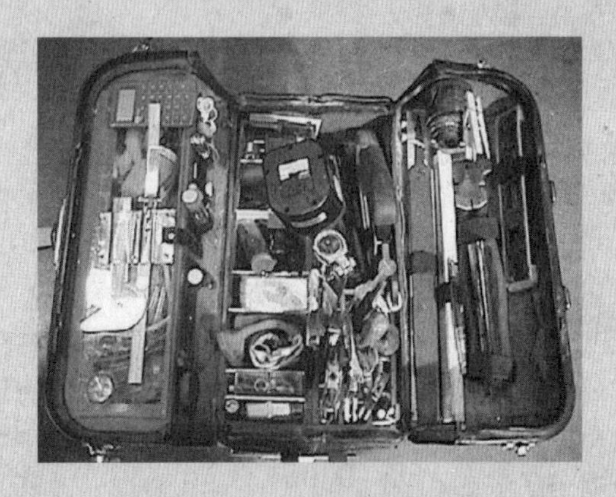

Much like the handheld digital camera with which I took these pictures . . .

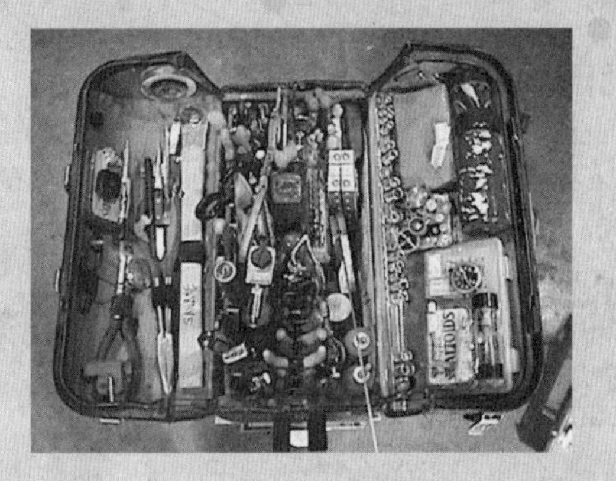

. . . the design and utility of my doctor's bags would evolve over time.

You can do almost anything with these toolboxes . . .

. . . except take them through a TSA security checkpoint.

that is both revelatory of your own beliefs about making and evolutionary with respect to how you work. And yet, here's the thing about evolution: it is not actually incremental. Yes, it moves slowly for a while, but then something big happens, something groundbreaking, and that's when the big evolutionary leaps that we can see with our eyes tend to occur.

As a maker, I took that leap once I stepped back from my aluminum doctor's bags and my deconstructed Kennedy stacks on rolling ladder racks and I realized that everything I was looking at in my shop was a reflection of my philosophy on making. Everything I used the most, everything I *loved* the most, was laid out in a way that increased its visibility or its accessibility, and was stored on something I either bought or built with an eye toward speed and agility. Everything I used least, everything that made me nuts, was always something that slowed me down. That understanding produced a leap in my creative consciousness that got me started trying to turn as much of what drove me batty into something that drove me wild.

I began taking time to reflect on my work as a way to monitor my usage patterns. I watched how I grab things, and at what kind of interval I need them. I am continually looking for ways to adjust the layout of the shop to be more efficient, that is, to increase visibility and accessibility. Probably 20 percent of what I do on a daily basis, in fact, is making small changes to how things are organized—piece by piece, bit by bit, shelf by shelf, drawer by reluctant drawer.

My shop is, of course, tuned to my needs. It is a reflection of me and a manifestation of my philosophy on how to work. Visual cacophony and first-order retrievability are just phrases I came up with to help me communicate what I believe to my collaborators, to my team members in The Cave, and in darker times, to myself.

They are not gospel. Your beliefs will be and should be your own. Your answers to the big questions will be different. Maybe, like Nick Offerman, you work best when things flow along the natural continuum of whatever it is that you make. Or maybe, like Jamie, your mind is most creative and your hands are most productive when everything is compartmentalized and in its place.

Or maybe you're nothing like any of us. Just never stop figuring out the most creative and productive ways to refine, and send pictures. Because I'm always looking for a better way . . . usually, as fast as possible!

CARDBOARD

While the things I make tend to fall into a handful of categories like replica film props and cosplay costumes, with one look inside The Cave you'll see the tools, techniques, and materials of a generalist maker. There are saws for every surface, cloth for every type of costume.

Like my choice of projects, though, there are still materials that I love and prefer over others. While I enjoy working on finicky, specialized stuff like leather or fiberglass or acrylic, I have the most affection for materials whose uses I can clearly identify and whose limits I know how far I can push. These allow me to make mistakes while I endeavor to understand a particular project better, and in so doing become conduits for learning how to work with other, more complex materials where trial and error is less forgiving to my watch and wallet.

In my realm, cardboard is king and the gateway drug to making.

Cardboard has a lot going for it. It is super cheap, it's easy to get, and it's easy to work with. More than 50 percent air, cardboard is rigid for its weight, which makes it easy to cut and easy to assemble with a great range of solutions: from masking tape to paper rivets, from hot glue to PVA glue, to all-purpose household cements, even contact cement. This makes cardboard an excellent training ground for all future material exploration. And once you can make a variety of things with cardboard, you're on your way to knowing the basics of sewing, carpentry, and welding, to name just a few skills, because all of these processes are the same whether you're making things out of leather, wood, sheet metal, or cardboard: it's just planar forms being joined under specific rules and conditions.

Laminate several sheets together with some fiberglass, for instance, and you can make airplanes with it. In the film industry, it's regularly used for mock-ups: quick and dirty builds, both giant sets, and hand props. I use cardboard constantly to try out designs for size, to get a real-world feel for what I'm doing, and to take my more complex ideas to an intermediate, risk-free stage between conception and drawing on one side and fabrication and assembly on the other.

I have been exploring my secret thrills and giving physical form to my creative obsessions through cardboard since I was about eleven or twelve years old and I came across my first abandoned refrigerator box on my way home from school. Nearly twice my size, it screamed out to me with the possibilities inherent in its wide planes of pristine cardboard, and so I started the process of pushing it home. It took me over an hour, and that hour was not without its difficulties. About halfway home, I came across a local bully, Peter. He could see that I was involved in something that mattered to me, which to a bully is like a red cape to a bull, so he decided to stop me.

Now, I'm not a fighter. I've only ever been in a single real fight, and I lost it handily. Likewise, I am very conflict averse, but that doesn't mean I'm not willing to stand my ground. Peter did that thing that bullies do, which is to assert that whatever it is you found or have is actually theirs, like they own the whole universe. Peter insisted this was *his* refrigerator box. I strenuously disagreed. He got in my face to further his point. I responded by pushing him backwards, to assert both my physical space and ownership of this box that I'd just spent half an hour dragging home. Peter responded to my response by screaming, "Didn't move me! Didn't move me!" as if the laws of physics no longer applied. Not only did he move when I pushed him back, he then moved clear out of my way. I would like to believe that my courageous display of fortitude made him reevaluate me as a peer, but it is far more likely that he realized getting into an actual physical confrontation over a giant box was probably not the best use of prime bullying hours.

When I finally arrived home with my prize, I placed it on the front porch triumphantly, while I thought about what I'd want to build. So many possibilities! This, I think, is why children and makers love cardboard more than anyone except maybe Jeff Bezos: the limitless options offered by this simple, easily available, surprisingly structural material. To a kid, a box can be a car, it can be a shelter, an escape vehicle. It can be a super-smart computer assistant, a playmate, or a Death Star. Or in the case of this box I'd just found, a spaceship.

My friends, the Caro twins, were making a Super-8 space-based film. When I told them about the box, we determined that I could make a viable spaceship cockpit out of it. So that's what I did. We filmed over a couple of days, and when we were done, I had a spaceship cockpit! I promptly installed that cockpit in my parents' guest bedroom closet. I left about two feet of space in front of the cockpit "windows" and painted that part of the closet with

black paint (without permission!) and then painted a star field in white acrylic. The previous years' Christmas lights supplied light and when I closed the door, I was in a freaking SPACESHIP that I HAD BUILT.

Two years later, it was in making with cardboard once again that I first encountered this sense of deep serenity. At the time I couldn't explain the sensation—I could only follow it—but years later I would come to learn that psychologists often refer to it as the "Oceanic Feeling," a oneness with the universe. I was about fourteen years old, and I had decided, after a bunch of paper modeling, to go big and make myself a cardboard man out of *another* refrigerator box I'd dragged home through the neighborhood.

I cannot explain why I made what I did, but I knew instantly that in its construction I had encountered something truly special. It felt like I'd scratched the surface of something important. I was, in one moment, completely overcome with this very fascinating feeling. It was neither ecstatic nor frightening. The best way I can describe it is of feeling both infinitesimal and universal at the same time. As if I were both the smallest thing in a room at the same time as being the *room itself.* (I promise, I wasn't high.) The tension of those two competing, yet covalent states, was invigorating.

When I was done with my cardboard man, I remember distinctly going into the kitchen and telling my mom: "Mom, I wanted to let you know that on this day, in September 1981, I am truly happy."

Later that year, still floating on the waves of this oceanic feeling, my art teacher, Mr. Benton, gave us an assignment to make something, anything, out of corrugated cardboard. Having already made

OPPOSITE: Why I decided to make an adult man with a big beard, dressed in a suit, that I sat out on our front porch, is a question that is beyond me to answer, even now.

a few large things from normal cardboard, I came home and excitedly mentioned the assignment to my dad. He took me to his local art store in White Plains where, with great ceremony, he bought me ten sheets of perfect 36-by-48-inch corrugated cardboard.

These pristine cardboard sheets were something new entirely. Their wide, flawless expanses sparked a level of inspiration that stood a degree beyond the beat-up refrigerator boxes from which I made my spaceship and cardboard man. I found the material so evocative that where the assignment asked for a single object, I made something like nine or ten. I simply couldn't stop. I made an MTV logo (the fledgling music network had just begun on cable), I made a full-sized electric bass, with four strings made from twine. I built a turntable where every edge was finished with the outer skin of cardboard so that when I was done, it looked like a turntable someone had painted brown. I can still recall the fever with which I dove into the material, again and again, making new objects until it was time for bed.

Working with cardboard was the first time I fully understood how a simple, inexpensive, ubiquitous, versatile material could contain limitless depths. It provided meaningful experience, endless creative possibility, and true joy. Plumbing those depths trained me for a lifetime of making. The lessons I learned working with cardboard informed all of my successes at incorporating new methods and new materials into my repertoire. It's allowed me to ideate, problem solve, communicate my ideas, and to feed my need to make stuff all the time.

MOCKING UP AND COMMUNICATING IDEAS

As makers, we become adept at holding complex concepts and ideas in our heads that we have our own mental shorthand for. This isn't a problem when we are working for ourselves on projects of our own. But when we're working for someone or with someone, quite often this tendency toward internal cataloging can lead to many of the fundamentals to a project going unspoken because they've become baked into our assumptions about what everyone should already know. But it's important to remember that this kind of knowledge is not so easy for civilians to engage with and process.

In my professional life, I have worked with every conceivable type of client and collaborator, from those who were makers with the same or greater expertise as me, who understood deeply what I was talking about when we discussed a build, to clients who couldn't glue two blocks of wood together if you put the blocks in their hands, covered with glue, and told them to clap. Being able to communicate your ideas to clients and collaborators is one of the most important skills to possess as a maker, otherwise some of your projects may never get off the ground.

One of the best ways to bridge that communication gap is actually not with communication at all. Or at least not with verbal communication. In the making of three-dimensional objects, it's ludicrously difficult to communicate complex shapes (and often even simple shapes) using only words and hand gestures. Just as writing a book involves an outline and a rough draft (so many drafts!), which get polished into a final manuscript, making things often benefits from a preliminary stage where the big details get worked out, and then a final fabrication stage where the small details get worked out. Cardboard is a low-threshold material that can make discussion of ideas at the preliminary stage so much easier and more complete.

I learned a long time ago that the best way to begin any job is to start with a rough mock-up. This has been an important tool for me, and a completely indispensable one for talking to clients

These Terminator 3 *sets were actually completed between the time Jamie and I shot the pilot for* MythBusters *and when the first show premiered.*

and collaborators. It allows them to see the size, the shape, the scope, and the scale of what I am going to build for or with them. In model making, where I've done most of my work over the years, it's hard to beat cardboard as the material of choice for these mock-ups. That's especially true for special effects sequences in movies, where directors of big sci-fi and fantasy films will ask for large miniature sets (or "bigatures" as Richard Taylor and our friends at Weta Workshop dubbed them after the colossal miniature sets they made for the *Lord of the Rings* movies) to be mocked up before they build and shoot the real model.

I've built many of these cardboard mock-ups, including one for *Terminator 3* that featured a character walking past a wall of windows on the other side of which is the entrance to a particle accelerator. Before embarking on the filming miniature, which would be six feet wide by eight feet long, my friend Fon Davis and I quickly built a ¼ filming scale model in about two and a

That's Fon Davis rigging the filming set for shooting on the motion control stage.

half days. Seven weeks after we made the model, we produced the filming set. It was a beautiful set, and the last job I completed at Industrial Light & Magic before moving on to *MythBusters*.

The purpose of these cardboard mock-ups is twofold: to give the film crew a sense for what they'll be working with and what kind of issues might arise during photography: tough shooting angles, limited camera movements, things like that. But also to give the model makers a quick and dirty way to get a feel for what we're getting ourselves into—to solve small issues before they become big problems during fabrication; to communicate with various members of my team what they will be working on and how it fits into the whole. Heaven knows, in my early days at ILM, having that broader understanding of my purpose made it easier to endure the tedium of cutting out and applying a thousand identical space shuttle tiles or etching twelve stories of a continuous wraparound staircase on a rocket gantry.

ARCHITECTURAL MODELING AND STRUCTURAL PROBLEM SOLVING

When my wife and I bought our first house together, one of the very first things I did was to measure every single room and build a 1/24th cardboard scale model of the house and the lot it sat on, including our backyard. The primary reason I did it was to get to know the house. By building my house in scale, and giving myself a bird's-eye view of my shelter, I was putting the totality of it into my head, and into my body. It taught me things about it I might not have learned otherwise. For instance, I figured out how to move the HVAC system into a better spot on the second floor by finding a blank spot behind a closet. It also became an invaluable way to talk to our landscapers about where to plant trees, our architect about building a deck out behind the house, and each other about where to put various pieces of furniture. (It helped that I built scale models of most of the furniture in the house out of cardboard, as well.)

When we moved on up and bought our second house several years later, again the first thing I did was to build an architectural model of it. The architect we hired to renovate the ground floor was both humored and impressed by it and it became an indispensable tool for discussing options.

Making a cardboard model of your living space or your work space is super fun. It gives you a fantastic perspective of where you are, and it's easy. I usually start by taking a paper bag, or a piece of butcher paper, and I draw out a rough floor plan of the space. I draw in arrows for the dimensions I'll need, and I place a small box in the middle of that arrow. This makes it easy to see if I've missed any specific measurements.

Then I move through the house filling in each box until I have all the basic measurements (in inches) that also answer more func-

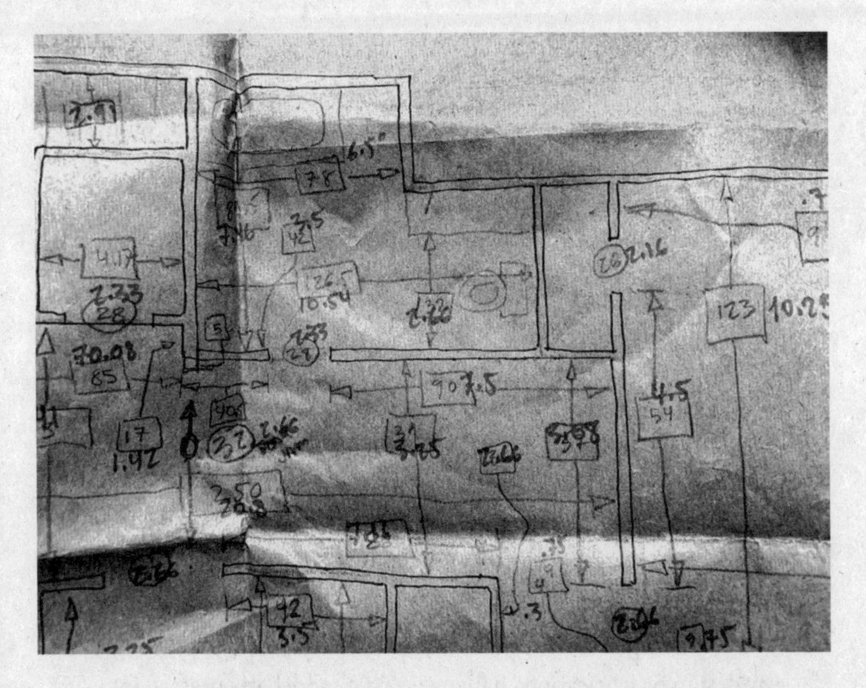

This plan is done in 1/12 scale (1" = 1'), which is called dollhouse scale. It's a nice big scale, but trying to build your whole house at 1/12 can get unruly, so in general I build rooms in 1/12 scale, but for entire floors of a house I go to 1/24 scale.

tional questions about the space itself: How far is the door from the end of the wall? What is the spacing between the windows? All of that goes down. Now that I have the actual measurements, it's time to convert to my scale. I move through box by box dividing by my scale number (usually 12). I add those new scaled values in red, to separate them from the visual cacophony the drawing has attained at this point.

What I'll do then is draw out my measured plan of the house in proper scale on a piece of cardboard, leaving a border around the perimeter to give the model some breathing room and to give me some cushion to erect the walls, as most plans don't account for the actual thickness of walls between rooms. Then I cut a whole bunch of cardboard strips that are the scaled heights of my walls, and start to assemble the model with hot glue.

I've done this enough now that I can do a whole floor of a house in about three hours. It might take you a little longer to build a scale model like this, but you should still try it, nonetheless. You'll be surprised how much better you understand your physical space after you've processed its dimensions through your mind and your hands. And the amount of time you save when it comes to doing the building or making the decisions that were the whole reason you built the model in the first place will blow your mind.

On *MythBusters*, we did a few stories using a fifty-foot-tall fire tower in Santa Rosa, California. I made an architectural model of it out of corrugated cardboard, so Jamie and I could work through any design or fabrication challenges. That model was pressed into service every time a build involved welding something to the top of the tower, because bending a lot of coat hangers and gluing them to cardboard to simulate whatever steel rig we needed, was a lot easier than traveling to Santa Rosa and climbing a five-story tower with a bunch of heavy gear to do it on-site.

This is the value not just of scale models, but of light, durable, versatile materials like cardboard. And by the way, while we're talking about cheap and easy-to-find materials, the humble coat hanger deserves a little more attention. I love coat hangers. I love them so much that every single bugout bag I've put together has had a few lengths of coat hanger wire in it. I've used them to open locked cars, scratch my back, and to build countless scale rigs over the years. I've made whirligigs, and also durable costume parts out of them.

The coat hanger is such a perfect piece of wire. It's rigid enough to be actually mechanically useful, yet soft enough to be cut easily with almost any tool you might have. It's very bendable, but holds its shape. It's also everywhere: In life, one is never too far from a wire coat hanger. In my shop, I've concretized that ubiquity by dedicating a specific rack just for coat hanger wire. A week does not go by without that rack getting rolled over to the workbench.

PROP AND COSTUME PROTOTYPING

My friend Max Landis, who is a screenwriter and showrunner, asked if I wanted to build a fully functional hero weapon for a new show he was working on. It was based on a Douglas Adams's book called *Dirk Gently's Holistic Detective Agency.* He was looking for a steampunk crossbow Taser type of weapon that would fit the style of the sci-fi show. This, of course, was right up my alley. Initially my plan was to use cardboard for the model prototype, but the rounded edges that a piece of weaponry like this required meant I would have to lean on another of my favorite lightweight, versatile materials: foamboard. I spent a full day cutting and carving and shaping pieces of foamboard into the weapon's constituent parts: the bow, the handle and trigger guard, the pistol grip. I labeled every single component with the materials from which I thought each part should be fabricated once Max handed off the prototype to his art department.

Now, in theory, I could have made the whole thing myself, with all the materials I was suggesting to Max. But that would have taken an incredibly long time to get right. Not just sourcing the materials themselves, but working with them to see how they fit together, going through the roller coaster of trial and error that is inherent to the prototyping process. It would have also been

incredibly expensive. And that was not what either of us were looking for. Instead, the foamboard prototype gave Max an awesome hero prop (if I don't say so myself) that he could task his team with building; and I got the thrill of creating it for a friend, one that shoots arrows AND lasers!

Beyond props, my earliest costume work was in simple material like cardboard and foamboard, as well. Remember that Excalibur armor I made out of aluminum sheeting and pop rivets as a Halloween costume my junior year of high school? That was actually second-generation armor. The first generation was two years earlier, when I built a suit of armor AND A HORSE entirely out of corrugated cardboard.

In that instance, as a fourteen-year-old, cardboard was my material of choice because it was my only choice. Today, I can make armor out of anything I want. Others aren't as lucky as I am in that regard. Fortunately, over the past few years, the advent of digital technology has fomented a sea change in cosplay circles that has opened the floodgates for makers and cosplayers with limited means.

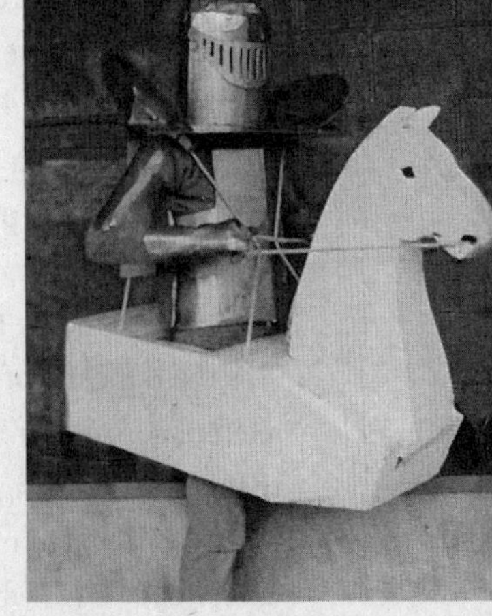

A program called Pepakura can take 3-D drawings and "unfold" them until they're printable as templates that fit on tiled sheets of standard copy paper. These templates are often used for carving pieces out of EVA foam (like camping mats) but can just as easily be transferred to cardboard, with stunning results. It's a great way to participate in the ritual of cosplay culture, if you have any interest, as well as to gain invaluable experience with costume design and construction. There's no better way to understand how planar forms can join together to create all varieties of shapes than by following a template and using a material that doesn't cost you an arm and a leg to work with.

CROWN YOUR KING

To make anything, it's critical to have a physical understanding of how all the component parts of your project will fit together. For a maker like me—a maker of larger physical objects like props or costumes or filming models—cardboard is king. It is a great, super-cheap way to develop physical understanding. With a matte knife and a hot glue gun, you can quite simply slap together enough of your design to understand how it might feel in the hand, how it might relate to other items it will interact with, and how many problems you might encounter as you go about fabrication and assembly.

But maybe cardboard isn't for you. Maybe it's not the best material for what you make. That's completely fine. Just figure out what material works best for you to experiment, to mock up ideas, to prototype designs, and to make mistakes without putting yourself behind the eight ball, either in terms of time or money or momentum.

For Tom Sachs, that material is what he calls his "sacred scraps." They are pieces left over from previous projects or earlier phases of a current project that were the by-product of actual work. "It's kind of the negative," is how Tom described it to me. "It's the opposite of the thing that you made, so you already have a template. You already have something that fits exactly the thing that you're working on. It might be the thing that you need, for example, if you're making a shim, or if you cut a piece too small, you already have what you need to extend it."

Even better for Tom, when it comes to experimenting with solutions or messing around with different, yet related ideas to other things he's made, sacred scraps often bear the markings of those projects. "They are the vestige of earlier projects, maybe ones that you didn't even remember," Tom said, "but you had pencil marks, you had screws driven into it, maybe nail holes, maybe a factory edge on one side, maybe a cut edge on another, maybe it was painted, maybe it wasn't. . . . And you had all these random but authentically generated representations of experience; in other words, these things had a past. It wasn't new material, so it had soul and history." A history he could use as a guide he could lean into as he played around with his ideas.

For Andrew Stanton, as a writer, his version of cardboard is the laptop. Until the laptop came around, Andrew hadn't really ever considered writing as something he could do. He'd been animating and storyboarding and working primarily on the production side for most of his first years at Pixar. But when Pixar got the green light to write and produce its own film and Joss Whedon showed him that writing movies was really, in his words, just "cinematic dictation," he saw how it could be done on the laptop. It was translating the movies we see in our heads into words on a page.

"Well, I can do that. I've had movies in my head since I was a little kid," Andrew told me. "And then the next half of the equation that got me over the hump was the invention of the laptop. I suddenly had no fear of writing on a word processor. Because it was sculpture. It was literally asking me to just spit crap onto a screen and then cut and paste it. The process invited messiness. It allowed slop. And I could refine later. Because of my upbringing, and I'm still like this, if I put a pen to a piece of actual physical paper it's got to be better. It's got to be an essay; it's got to be good; it's got to sing on the page. And that stops me cold. The laptop helped me realize, oh, I can be messy."

What material can you wrap your arms around to gain a complete sense for the skills you want to master and the objects you want to make? Is it cardboard? Muslin fabric? Crappy butcher cuts? Scrap wood? The backside of recycled printer paper? A word processor? It really doesn't matter, as long as it allows you to be messy and it keeps you moving forward in your journey as a maker.

HAMMERS, BLADES, AND SCISSORS

Humans are toolmakers. We are explorers, innovators, inventors, and what facilitates all of that is our use of tools. I feel like the hammer *must* have been the first tool: a rock to crack something open, or to drive a stake into the ground. One swing to subdue dinner, or an enemy. The hammer is the ur-tool. Similar to early man, beginning makers start with a rudimentary set of tools for basic creative tasks: a hammer (of course), a set of screwdrivers, scissors, some pliers, maybe a crescent wrench, and some kind of cutting device. Almost everyone who has strived to make things has some combination of this list. Then, as we get more experienced, we seek out better versions of the tools we already have as well as new tools that can facilitate the learning of new techniques—new ways of cutting things apart, and new ways of putting them back together.

Once we start to expand past the basic complement of tools, what to add to our collections becomes a multifactor calculus based on reliability, cost, space, time, repairability, skill, and need.

These choices are nontrivial, because the tools we use are extensions of our hands and our minds. The best tools "wear in" to fit you based on how you use them, they get smooth where you grab them. They tell the story of their utility with their patina of use. A toolbox of tools you know well and use lovingly is a magnificent thing.

But how do you get there? How does one begin to build such a collection? This is one of the most common sets of questions I get about the practice of making. Budding actors obsess about "process," budding writers obsess about routine, budding makers obsess about tools. Like being more worried about how to dress for an interview than how to answer the questions you'll be asked, each imagines that the magic lies somewhere in the approach to the act of making, as opposed to the making itself.

The reality is that tool choice is both less important and more important than you think it is. It is less important to the extent that tool usage is entirely subjective, which means there is no one right way to do things. But it is more important, because the best tool for any job is the one you're most comfortable with, the one that you can make do what you want it to do, whose movements you fully understand. For me, of course, the tool I'm most comfortable with is the Leatherman, which is why I call it my third hand and use it for everything—from tightening a drawer handle, to clipping a hangnail, to banging a door hinge back into place, to picking gum out of the treads of my shoe.

I learned this lesson about tool choice and usage from my friend Mark Buck, whom I worked with at ILM. Mark had a round, kind face and an awesome goatee that belied a fine mind, a biting wit, and amazing building skills. He was a man with an aversion to fools and a love for tools. If he walked by your workbench and he saw you looking too long for something in your toolbox, he'd instruct you in an almost solemn tone, like a *sifu*, "Remember, in every tool, there is a hammer."* What he meant was that every tool can be used for a purpose for which it wasn't intended, including the most basic of operations, like hammering. He also meant that until you learn to see what tools can do beyond their stated purpose, you can't quite be a maker. Truer words have rarely been spoken.

My goal here is to carry on Mark's legacy and help you get comfortable with your tools while offering some guidance on how to fill out your tool collection as you get more experienced. It's not complicated, but it's worthy of some deliberation and some care, so you don't end up with piles of junk on your floor like I did when I was young. The key is to recognize where you are in the evolution of your skill set and your usage of particular tools or techniques, and to buy accordingly.

ON THE CHEAP

When I started out, like most aspiring makers, I built my tool kit with castoffs from my dad's studio and cheap tools I found while rummaging around New York City. That was all I could afford at the beginning of my career, and I suspect circumstances are the same for most makers in their early days. I got by on this mishmash—for longer than I expected, to be honest—but eventually, as I started to pursue actual *job* jobs at effects houses in Manhattan and theater companies in San Francisco, I had to start building up a stock of better tools.

My problem wasn't choice. There is an incredible array of choices for nearly every tool you could possibly want. The prob-

* While most manufacturer's warranties would object to Mark's advice, his wisdom was true enough to stick in my brain for twenty years and become the title of this book.

lem was that good tools cost good money, and there's nothing worse than paying hundreds of dollars for a tool that never gets used. It's throwing money away. So the first thing you have to do when you're building out your toolbox is figure out if you really need something. Whether you're thinking of adding a set of screwdrivers or a reciprocating saw, buy the cheapest version you can find. Don't just look at a discount hardware supplier like Harbor Freight, either. Look on Craigslist, go to garage sales, borrow from friends or your local maker space, beg your mom and dad to let you have what they're not using. Then, once you have a new tool you think you need, spend some time getting to know it physically. With certain tools, I'll go so far as to take them apart, just to understand them better, inside and out. I'll take that opportunity to make sure they're cleaned, and oiled, and properly tensioned wherever they need to be. If you are unfamiliar with a tool or inexperienced with the techniques required to use it, getting comfortable like this is the most important thing you can do, because you might really need this thing, but if you are intimidated by it, you aren't going to want to use it, and then what's the point?

Once you've got a new tool setup, it's time to put it into action. For some people, the newness will produce enough enthusiasm that they will find reasons to use it. For others, it won't automatically occur to them how to incorporate a new tool into the flow of the way they make things. If you're one of those makers, the only way you're going to get familiar with your new tools is to run them as a subroutine on top of your normal making patterns and problem-solving algorithms. And if even that doesn't work, and you still find it's hard to integrate a tool into your workflow, that probably means you don't need it. No harm, no foul. At worst you're only out the small amount of money you paid to conduct the experiment. If the tool turns out to be fantastic and indispensable, however, and you're able to integrate it seamlessly into the way you work, well then you've hit a home run and found a new member of your toolbox.

GET THE GOOD STUFF

Because I worked in film as a model maker, my toolbox was a professional expense, and I treated it as such. My job required me to be fast and efficient, so anything that helped me do my job better was a worthwhile expenditure. My personal rule was that if I needed a tool more than three times within a year, it was worth investing in a good one of my own.

Still, I always started with a cheap version, partly out of frugality, but also because I found that it helped me shop for the good one. If you've never used a tool before, reviews and articles about it can only get you so far. You need to work with a tool in order to see how the tool works for YOU. You need on-the-ground experience with it in your hands. If it ends up working out, and you decide to get a higher-quality version for yourself, now you know what you're looking for, you know what features you value and which you don't care about. It makes you a better consumer.

When I remade my old doctor's bag toolboxes out of aluminum and pop rivets back in the early ILM days, to that point I had not used rivets a lot professionally, so I had no use for a fancy riveter. Instead, I used an old hand-operated rivet-setting tool I had laying around. It worked just fine, but for one small issue: rivets can be hard on you, physically, and this tool turned my hands into crippled hunks of meat. Each one of the toolboxes required more than three hundred rivets and I had committed myself to finishing both of them in time for work the next day.

That meant hand securing at least six hundred rivets in the space of twelve hours.

When I showed up at ILM the next morning with my two beautiful toolboxes, I was very pleased with the attention they got. Then I tried to pick up a pencil to start making my to-do list for the day. I couldn't do it. I couldn't hold one for more than a second. The constant cranking of the rivet-setting tool had rendered my hand almost completely useless. Commiserating with one of my colleagues, he mentioned that, you know Adam, pneumatic riveters are actually a thing that exists that he definitely would have used if he were me. Unfortunately, I was not him, because somehow I had never heard of this tool. Then he pulled one out from the back of the model shop and showed it to me. I immediately wanted one very badly, but as I did the research I learned that they were a fairly expensive bit of kit. A good one was well over $200, which was real money, and having done only one big project with rivets—a personal one, no less—I didn't feel like I had a justification for the purchase.

Then one day, flipping through a Harbor Freight catalog, I saw a pneumatic riveter for twenty-five dollars. I ordered it immediately. It was worth taking a flyer on, in my opinion, because I've always loved working with rivets, going way back to high school and my Excalibur suit of armor, and there was a good chance other rivet-related projects would come along that required the tool.

I ended up being right. I used that twenty-five-dollar riveter a few times before it broke. It only lasted about three months (you really do get what you pay for), but that was long enough to know that a pneumatic riveter integrated beautifully into my workflow. Thus, when it failed, I had no issues investing in the expensive high-end version. I knew I'd use it. When you know what you're working with, investment in high-quality tools pays dividends on many fronts. Of course, they last a lot longer than cheap tools (I'm still using that same pneumatic riveter nearly twenty years later), but they also go out of alignment less frequently, they're easier to fix, they perform more true, and perhaps most importantly of all, they often just feel better when you're using them.

DIVERSIFY AND MULTIPLY

Once you're rolling and getting good at using the tools in your collection, it'll start to dawn on you that you might need more than one version of a tool. Don't misunderstand, I don't mean more than one *copy* of a tool—though that's perfectly reasonable if your personal philosophy of making includes first-order retrievability, like mine does—I mean more than one style or variety of the same basic tool. There's a whole world of hammers out there, for instance. You could spend your life buying every single variant of needle-nosed pliers. And chisels? It feels like there are more types of chisels than there are wood species to carve.

These varietals exist not to break your bank, though they will often cost a pretty penny, but rather for specific, and sometimes single, use cases. Let me give you an example. One day working in Jamie's shop in the early '90s, I needed to cut a sheet of acrylic plastic. Acrylic is a very shatterable plastic, and while you *can* cut it with a regular table saw blade, the fact is, you shouldn't. Your edge will be rough, the piece might not make it all the way through without utterly shattering (I've seen it happen), and it's just bad practice. You want a nice clean edge for acrylic, and there's a specific blade designed to do just that. It's usually got what's called a "zero kerf" (the sides of the teeth on the blade don't stick out over the plane of the blade itself) and "triple chip" (the teeth proceed around the edge of the blade with one tooth sharpened slightly to lean left, the next one to center, and the third slightly to the right). I'm not super sure why this blade in particular is the best for cutting acrylic and leaving a super clean edge, but trust me it is. And it is expensive for that very reason.

If you are a maker who cuts acrylic every so often, it makes sense to have a blade like this and it would also make sense to use it properly and be careful with it. But that's easier said than done in a shop that has a lot of people working in it. Invariably, someone will pull out the acrylic blade to cut some plastic and forget to put it back. The next person at the table saw might rip a huge sheet of plywood, and then the third person might cut a chunk of aluminum, neither of them aware of which blade is in the rig. Before you know it, your expensive zero kerf, triple chip plastic blade is garbage. This happened in Jamie's shop more than once, inspiring someone to leave a note taped to the shelf where the plastic blades were stored:

> Hey weasel-wort! Yes you! I hope your cut goes really well but please, for the sake of all that is holy put this damned blade back in its protective case when you're done. Don't be an a-hole.

Years later, when I invested in my own zero kerf triple chip wonder blade, and it cost my own hard-earned money, I wrote simply:

> For plastic only!

It sincerely felt like I was maturing as a maker and shop proprietor to have an expensive blade like that, one that was meant only for one thing. At the same time, it also felt a little excessive and maybe a little precious. Do we really *need* all these different saw blades? Can't we get away with maybe fewer than twenty-three different types of hammers? The answer, of course, is yes, but it wasn't until I talked to Kevin Kelly, the legendary founding editor of *Wired* magazine and a tool enthusiast, that I truly understood why, as makers, we should embrace the diversity, and why I, personally, have so many tools.

"Freeman Dyson, a famous physicist, suggested that science moves forward by inventing new tools," Kevin began as we talked on the phone one morning all about tools. "When we invented the telescope, suddenly we had astrophysicists, and astronomy, and we moved forward. The invention of the microscope opened up the small world of biology to us. In a broad sense, science moves forward by inventing tools, because when you have those tools they give you a new way of thinking."

Kevin was putting words to how I felt whenever I got a new, unique tool and used it for the first time. It always made the task for which it was designed so much easier than it was with the more conventional tools, and that ease of use invigorated me and freed up my mind to think about what else I could do with this thing.

"What's happening with individuals engaged in craftsmanship is that when they change their tools, they have different ideas. It gives them a new different view on the world. It opens up a new possibility space. Tools are the way you explore possibility space, the space of possible things."

YES.

"In the beginning, when you're young, you think you have certain choices, but oftentimes, a new tool will open up a whole new space that you didn't even know about and the way you get through that space is through the mastery of those tools."

YES. YES!

"So then there's this sort of epiphany where, because you know what it's like to use the tool, it opens up all the things that you could do with it. All of a sudden, now you have a power. Now you have a possibility space that you can explore that you didn't even know about before," Kevin said, as I sat at my workbench in The Cave, staring over at a drawer labeled Scissors. Everything Kevin had just said spoke directly to the contents of that drawer.

Now, you might think that a shop only needs a single pair of scissors, or in my case three pairs of scissors distributed strategically around the shop, but, hoo boy, that couldn't be further from the truth. The fact is, there are dozens of different things you need distinctly different scissors for.

There's regular paper scissors that you can get at the hardware section of your grocery store. Those are fine. Buy them cheap, abuse them, buy them again. That's my policy.

There's also tin snips, for sheet metal. Your run-of-the-mill scissors can cut through foil well enough, but for anything thicker that's made of metal you need a good pair of tin snips. Just go ahead and invest in a good pair that feels right in your hands. And wear gloves when you use them because the "paper" cut you can get from metal will send you to the hospital. Trust me. I've been there.

After tin snips you should have a pair of medical scissors. These have tiny fine blades on them and are fantastic for getting into tight places. I have a pair in my first aid kit, and another on my model-making tool stand. They come in handy more than you'd think.

Then there are the thread cutters. They're for snipping the extra thread from that line you've just sewn. Using a big pair of scissors for that operation works, but these are purpose-built to be small, highly useful, and easy to maneuver. They are so much better that when you get them, you'll write me a thank-you note. They also come cheap, so go ahead and buy ten-packs online. You're welcome.

Last but not least, there are fabric scissors. A good pair of cloth-cutting scissors is the only other type of scissor a maker MUST have in their shop. Not only that, it is the one scissors that you must protect the way you do an acrylic blade. Specifically, you should never EVER EVER let anyone use your fabric scissors to cut paper.

Not many people know that cutting paper dulls steel blades faster than most things. The fibers and other materials that make up paper are, from a material science standpoint, terrible for a steel edge. Some coated papers contain up to 30 percent clay, which is an abrasive, and some recycled papers carry around much of the trash they were made from, including microscopic bits of metal. Suffice to say that cutting paper is terrible for any scissors, but especially fabric scissors. That's why in my shop, on every pair of fabric scissors we have, I write in bright White-Out marker: CLOTH OR DIE.

I've done this since 1995 when I caught a production assistant on a TV show I was working on using a one-hundred-dollar pair of my best fabric scissors to cut a fake flower stem. You might not know this (he certainly didn't) but inside a fake plastic flower stem is a thin rod of hardened spring steel piano wire that will put a permanent ding in a pair of fine scissors, rendering them worthless.

I take Kevin Kelly's point to heart about tools as gateways to new spaces and greater understanding. It can still feel like an indulgent luxury to have tools in your shop that do one thing, and one thing only. What you have to remember is that you did not begin at this place where you are buying acrylic saw blades instead of combination circular saw blades, or upholstery hammers instead of standard claw hammers, or fabric scissors instead of cheap-o all-purpose scissors. You arrived here as a consequence of meaningful experience. You've made enough stuff—including mistakes—to know what your workflow patterns are and to know the benefit of having a tool that will do the job right on the first try.

THE SUPER-SECRET SPECIAL TOOLS

The last level of tool acquisition is one that you cannot reach by yourself. You must be lifted up to it by your maker community, by your collaborators and coworkers and clients. It is a class of tools—and I'm including techniques in this category as well—that you would never have known about but for someone more experienced sharing their knowledge with you.

I've had dozens of these tools shared with me over the years. There was a Japanese pull saw; these beautiful C-Thru rulers and triangles that have a metal edge for cutting that I have been using for thirty years; and Forstner bits that blew my mind when I discovered that such a tool existed. After someone introduced me to step drills, I remember telling an engineer friend, "Dude, I have a drill that can drill a one-inch hole through one-thirty-second-inch acrylic and not shatter it." And he was like, "No way!"

If a film crew is an army of problem solvers, a model shop is an army of inventors. One of the things we loved to do at the ILM model shop was to trade surprising and inventive tools, techniques, and tips. Good ones raced through the shop like electricity. I was the beneficiary of far more than I was ever the supplier of, but the one great one I remember sharing was a real crowd-pleaser.

I was working on *Space Cowboys*, and we had to make a Russian satellite that was almost ten feet tall and covered with a dozen radar dishes. I was tasked with making these dishes in miniature,

I made an extra radar dish for myself. Model makers do this a lot when they make things that tickle them. Shhhh, it's a secret.

each one meant to look like a steel truss frame, welded into a bowl shape.

It's easy to make something like this by gluing lots of little thin strips of styrene together into truss shapes, but that method is extremely time intensive and on this job I only had a couple of days, so I had to innovate.

I laser cut the framework out of a flat piece of one-millimeter-thick acrylic. Then, using the wood lathe, I carved a seven-inch-diameter bowl shape to use as a form, placed the bowl underneath a heating element, put my flat laser-cut framework on top of the bowl, and let the heating element heat the acrylic until it slumped into a bowl shape. With this method, I was able to make a dozen radar dishes in a couple of hours.

Lorne Peterson, one of the original model makers from *Star Wars*, came by my desk and, curious about what I was doing, asked me about it. He really liked the solution, and spread the word. By the end of the day most of my coworkers made it past my desk to take a look at this new technique and figure out how they could incorporate it into their repertoire.

Once, when a new model maker accidentally used steel wool on a giant clear vacuum-formed window that had been built for one of the buildings in the film, *A.I.*, none of us knew how he might fix it. Polishing microscopic scratches out of a large piece of acrylic (that needed optical clarity for camera) was almost an impossibility. As model makers we'd all been there. And the tragedy was that the form used to make the window had broken, so there was only one window, which meant he had no other option but to fix THAT window. His solution was as amazing as it was terrifying:

He put a liquid weld-bond glue that dissolves acrylic into a frying pan on a hot plate. Wearing a chemical respirator, he heated the glue until it started to vaporize, then held the window a short distance above the pan, letting the hot vaporized weld bond smooth out the microscopic scratches. It was *Breaking Bad*–level crazy, but it worked. It's one of those secret techniques that falls solidly into the category of Don't try this at home! Personally, I would never try it anywhere—the idea of heating up a noxious solvent genuinely scares me—but that doesn't mean we didn't all talk about it for weeks. And I'm sure some of my old ILM compatriots have whipped it out in a jam since then. That's just what you do when you're a maker and you've got a problem that demands a solution—you pull every arrow out of the quiver until you find the one that flies truest and hits the bull's-eye.

Here are a few additional arrows for your own quiver, courtesy of some of the great makers I had the privilege of speaking to over the course of writing this book. Some of these tools you may know, or even have, but each one is something that one of these great makers had never heard of before it was introduced to them, and it blew their minds once they gave it a try. Imagine how many tools like that are out there for you:

Jen Schachter, artist-in-residence at OpenWorks: digital calipers:

"I remember when someone first showed me digital calipers, that completely blew my mind, the three or four different ways you can measure things using them."

Bill Doran, the prop maker and foamsmith: a knife sharpener

"EVA foam dulls the knife really fast. I made the first *Mass Effect* armor using a utility knife to cut everything. And I'd do five cuts, throw away the blade, five cuts, throw away the blade, and I went through probably around one hundred blades to make that armor. It was slow, too. With a sharpener I just run the long blade across the surface really quick with a couple passes, then get back to making nice, clean shallow cuts in my foam."

Mark Frauenfelder, editor in chief of *Make:* magazine: head-mounted magnifier and the Big Squeeze:

"This ten-dollar head-mounted magnifier that I got on Amazon is something I never expected I would use all the time. You wear it like a headband and it has this big lens that flips down over your eyes and it's got layers of lenses on hinges so you can increase magnification.

"Also I use something called the Big Squeeze for a lot of stuff. It's a really nice cast aluminum tool with a kind of windup key. But it's not the windup that's cool, there are these two serrated cylinders and you run tubes through it—like paint tubes or a toothpaste tube—and it absolutely extracts every last molecule, anything that's inside the tube. It is the best. It's so good it will never break."

Nick Offerman, master woodworker and actor: the card scraper:

"The card scraper is a thin piece of steel. It's one of these things that you can buy ready-made but that you can literally make out of any scrap. Imagine a rectangle of steel that's three by five inches. It's like one-thirty-second of an inch thick. You put it in a vise and you take a screwdriver or a burnished steel rod and you lay it perpendicular on the long edge of the card, okay? You tilt it four to seven degrees to one side, press down and burnish the edge of the card creating just a tiny burr. So, you've got this rectangle of steel with a little burr. Then, using that burr as your leading edge, you hold the card in both hands with the burr on the bottom front, you bend it a little bit, giving it tension, and then you use that burr, pushing against your wood surface, to create shavings finer than a hand plane. This is the champion of this chapter."

I couldn't agree more.

SWEEP UP EVERY DAY

When my kids were still young and living at home, we bought a real tree for Christmas every year and decorated it together as a family. My job was always to put the lights on. A half-dozen strings of twinkly white lights that were, year after year, the absolute death of me. We'd come home from the tree lot, get our tree secured in its stand, then I'd open the box of decorations labeled LIGHTS from the previous year and be met with an impossible tangle of twisted, knotted cords and bulbs and plugs. Even if you don't celebrate Christmas, everyone knows this particular pain of unknotting a bunch of unruly cords. You don't want to take the hour it'll require to separate everything, but you know it has to be done. Some years were better than others, but only by chance, never because of anything I did.

Then one year, when it was time to take down the tree, I happened to have an empty mailing tube nearby and it gave me an idea. I grabbed the end of the lights at the top of the tree, held them to the tube, then I walked around the tree over and over, turning the tube and wrapping the lights around it like a yuletide barber's pole, until the entire six-string light snake was coiled perfectly and ready to be put back in its appointed decorations box. Then, I forgot all about it.

A year later, with the arrival of another Christmas, I pulled out all the decorations as usual, and when I opened the box of lights, I was met with the greatest surprise a tired working parent could ever wish for around the holidays: ORGANIZATION. There was my mailing tube light solution from the previous year, wrapped up neat and ready to unspool.

My initial reaction was shock—at both the ingenuity of the solution and the fact that I'd forgotten all about it—but that quickly resolved into gratitude. I was grateful to myself for taking the time a year before to save me this time now.

"Thank you, me-from-the-past!" I literally said to myself.

"You're welcome, future Adam!" past me called back from across the void.

But it wasn't just the organization of the Christmas lights that produced a sense of gratitude within me. It was the peace and the balance they represented in my life that generated the greatest swell of emotion, because order and serenity were not things that came naturally to me, as a man or as a maker.

Today, I am a fairly well-organized maker, but I used to be quite well known for the opposite. I was a messy guy. I mean like a REALLY messy guy. Those piles of trash on the floor of my first studio in Brooklyn, the ones that I sifted through to make art, that was nothing. Or rather, that was the tip of the iceberg. That lack of organizing structure carried through to my personal life, as well. I was a messy person, I was a messy roommate. I was awful to live with. I was a mess.

I was always too concerned with moving on to the next thing, too impatient to consider that taking the time to clean had both short-term *and* long-term value. I liked cleaning up when company was coming over, but aside from that I could (and did) tolerate a pretty rough style of living. At one point in Brooklyn, my cat, Regis, pushed a potted plant that was perched on my radiator ONTO MY HEAD while I was asleep. Dirt and plant stuff spilled all over my futon in the closet I called my bedroom. I brushed the dirt off the sheets and blanket and slept on that bed for two more weeks before I washed them. I was like that for a while.

I have since matriculated, slowly and methodically, into a neat person. I've embraced the mantle of a "clean" person. Nobody is more surprised than me, except perhaps my mother. In high school, my room was legendarily messy, with numerous models and costume constructions in various stages of completion marooned between massive cities made of LEGO bricks everywhere. Now that I have raised two boys to adulthood, I have come to understand that some of this messiness was simply adolescence, which is now recognized to last well into your twenties. For me it most certainly did.

But matriculating out of adolescence wasn't enough to spring me from the pigsty of my own making. Something else happened. Somewhere, something in me shifted and I changed from someone who never wanted to take the time to clean up, to someone who deeply enjoys organizing my work space.

FROM DIRTY TO CLEAN

The shift didn't happen all at once. It was gradual. It started in the second half of the 1990s, as I moved into my thirties, and began doing more serious freelancing out of the basement shop I had in my first apartment on Bartlett Street in the Mission District.

The Bartlett Street shop was the first raw space I truly designed to fit me and my needs. It's where I tried the first pass

Bartlett Street shop. Forty feet deep and twelve feet wide—I worked out of here for seven years, and supported my family out of this space for about four of those years.

of what would become my abiding shop philosophies and practices. Much of the time it was a craphole, but in its final two years I started to clean up before heading upstairs at the end of the day, and lo and behold the shop became a far more efficient and well-oiled machine to work in. The freed-up space in my mind and the open work space at my fingertips allowed me a lot of room, both mental and physical, to pursue a wide variety of projects, and I finally started to understand how much benefit was to be gained by taking the time to clean.

I gained even greater appreciation for sweeping up once I started on *MythBusters*, because now I couldn't do it as much, and

I knew what I was missing. When we were in full production, we filmed more than two hundred days per year. This meant that in a good month I'd get just a handful of hours in my shop, *total.* I curated my tool collection down to those that facilitated fast, accurate results, in the interest of maximizing what little time I had, but that often meant I'd be rushing out the door for filming, leaving the shop a mess that I wouldn't be able to get to for weeks.

When I'd have a week or more of concentrated time, taking care of and tackling the mess could sometimes take hours just to get back to a shop I could work in. It was completely untenable. I had to get back to spending my last few moments of the day sweeping up in the shop, however infrequently I was in there. The trade-off would massively benefit me in the momentum department.

By the time I moved into The Cave in 2011, that benefit was so clear, so disproportionate and unassailable, that it became as much of a pillar in my shop philosophy as the visual cacophony and first-order retrievability. To this day, I put away my tools and sweep up at the end of every day.

Of course, I don't always want to. There are plenty of times when a build has been kicking my ass until I can't look at it any longer, and I'm late to leave and I can't wait to get home where it's warm and quiet and there's not a set of failed attempts at solving a problem staring me in the face. But then I look at my shop and see the debris field that was that day's project: tools on every bench; extension cords everywhere; piles of bolts and screws; shavings of wood, or aluminum, or steel. And all I see, if that stuff is not cleaned up before I come in the next morning, is arrested momentum.

My entire organizational strategy, every second of sweeping and straightening I do at the end of each day, is about keeping up the momentum of my making. When I step into the shop each morning, the state of the shop will influence what happens in there. If the shop is messy, and I have to start the day just putting everything away, that has a small but appreciable effect upon my momentum. It starts me off the blocks with shoes made of lead. For someone who likes to work fast and use the momentum of iteration to bring out the best solutions in each of my complex builds, starting the day with cleaning is like starting off the Christmas tree decorating with a Gordian knot of lights. It's death.

But when the shop is clean, and I come in to an empty bench and racks full of tools where they should be, and I plop that day's project down and regard how to begin for the day, it feels like the air is electric with possibility. Grabbing that first tool feels so good, so right. Why would I not start every day that way if I could? It's fifteen or twenty minutes at the end of one day, in exchange for a fully productive six or eight or ten hours the following day.

This gets to the essence of the exchange I had with myself when I figured out how to store my Christmas lights: sweeping up every day and putting away my tools is a conversation between present me and future me. It is present me acknowledging that future me always likes to keep the momentum of a project going, and that having to look for a tool or walk across the shop to get one in the middle of a critical phase can slow down the creative process enough to be an existential threat to the project as a whole.

In his film *Ten Bullets*, about his own shop practices, Tom Sachs has a whole section on how to sweep. I mentioned my idea to Tom, that sweeping up is always a conversation between our present and future selves, and he pointed out that it is also an important time for reflection. He even thinks of sweeping as meditation:

"Well, I think what you're talking about, in a way, is a form of meditation or reflection because when you're sweeping up, you're looking at that pile of wood. You're connecting your thoughts with what you're making now, how that relates to what you've done, and how it relates to what you're doing next. And by doing that, you anchor yourself in the present moment, what you're physically doing, which is sweeping, but you're meditating on how your present thing connects with the past and the future. And that's called planning and reflection."

Indeed, planning and reflection are not just part of what it is to sweep up, they are essential to making as well. What are ideation and drawing and list making but planning modalities? What are checkboxes and experimentation and modeling but a reflection of what you have done and are trying to do?

FACE YOURSELF

I'm a fervent believer that a shop is a holy space wherein we pay attention, like prayer, to those things that are important to us: what we are doing, what we desire to make, how we solve the problems we encounter while tackling a project, and what we can achieve through our efforts. In each of those endeavors, to execute on them with rigor and excellence, we must, as with anything where the stakes are genuine, face ourselves.

By facing ourselves, I mean watching and learning from our own habits and making changes based on that information to improve ourselves. I mean facing down our own biases about how we *think* a project should go in favor of going along with where the project *wants* to go. Understanding this difference, between what a project should be and what it wants to be, and respecting the gulf between them, is key to fulfilling one's potential as a maker. It is a big part of what makes a master craftsperson.

Making at its most basic is a process of conception and construction. They are linked but not the same. We never conceive exactly what ends up getting built and we never build exactly what we conceive. Nothing ever goes exactly according to plan. Facing yourself means taking responsibility for that fact, and making peace with the reality that to build something real and substantive is to give up some measure of control over your preconceptions of what you imagined you were making in the first place.

To do that requires listening to some of the voices in your head and ignoring others: listening to the project, listening to the shop itself, and paying attention to its needs. When a shop is messy, it's noisy. A clean shop is quiet. It's a place for real reflection. And that's a good word to use, because I mean it both in the sense of meditation, like Tom did, and also actual reflection: the state of a shop is the state of its maker.

Any maker space (or recording studio, or drawing table, or sewing machine) is a place wherein the maker can safely experience the vicissitudes of life. We can screw up, and the stakes are far from life and death. We can triumph, and the crushing expectations of success do not immediately land on our shoulders. A shop is where I get to pretend that the universe has order and that I have some measure of control over it. Spending time prioritizing order and ease of use and cleanliness is projecting a value deep into your psyche.

I agree with Tom that it's a kind of meditation to clean up every day. And like the personal mantras of transcendental meditation, it has to be a meditation that works for you as a person and for what you do as a maker. Tom again:

"Spending time sweeping up and knolling at the end of the day is a way of reflecting on what you did, and putting everything away makes it really easy to start the next day. But I think it's important that everything is there for how it works for you, and that it's a pleasure."

There are few pleasures I now enjoy more than walking into

and working out of a clean shop. It's the rare day I leave the shop in a state that's less than perfectly clean. And yet, it still happens. I still run into days where the build was brutal, and I can't stand to be there one more minute. Sometimes I still need to bolt, to put an end to a crappy day and just unplug.

On those days, I don't beat myself up about it. I try to face myself, but I'm also trying to be gentle in the process. Present me has a quiet conversation with future me and promises that this isn't going to become a habit. The next day, a recharged future me usually responds that it's okay from time to time to abandon your best-laid plans and just get out.

It's okay for you, too. In the world of making, there's space for that. There's space for all of us.

ACKNOWLEDGMENTS

Writing this book was invigorating, educational, frustrating, inspiring, and harder than I ever imagined it could be. I couldn't have done it without the boundless generosity of so many. Drew Curtis, who started this weird ball rolling so long ago. Byrd Leavell, for getting it to fly. Nils Parker, for teaching me structure and flow and grace under pressure. Matthew Benjamin, for his love of books as important cultural objects and agents of change.

My early mentors also deserve credit for this tome. My father, for his advice as well as his example. Jamie Hyneman taught me what a boss could be. Mitch Romanauski taught me so much it's hard to quantify. Credit is also due to all the colleagues I've bumped elbows with at worktables throughout my career, too many to list but you know who you are. I want to give a shout-out to all of the young makers I am lucky to know who inspire me every day. For reals.

Finally, to my family, and especially to my wife for her seemingly limitless faith in me. And her biting and awesome sense of humor.

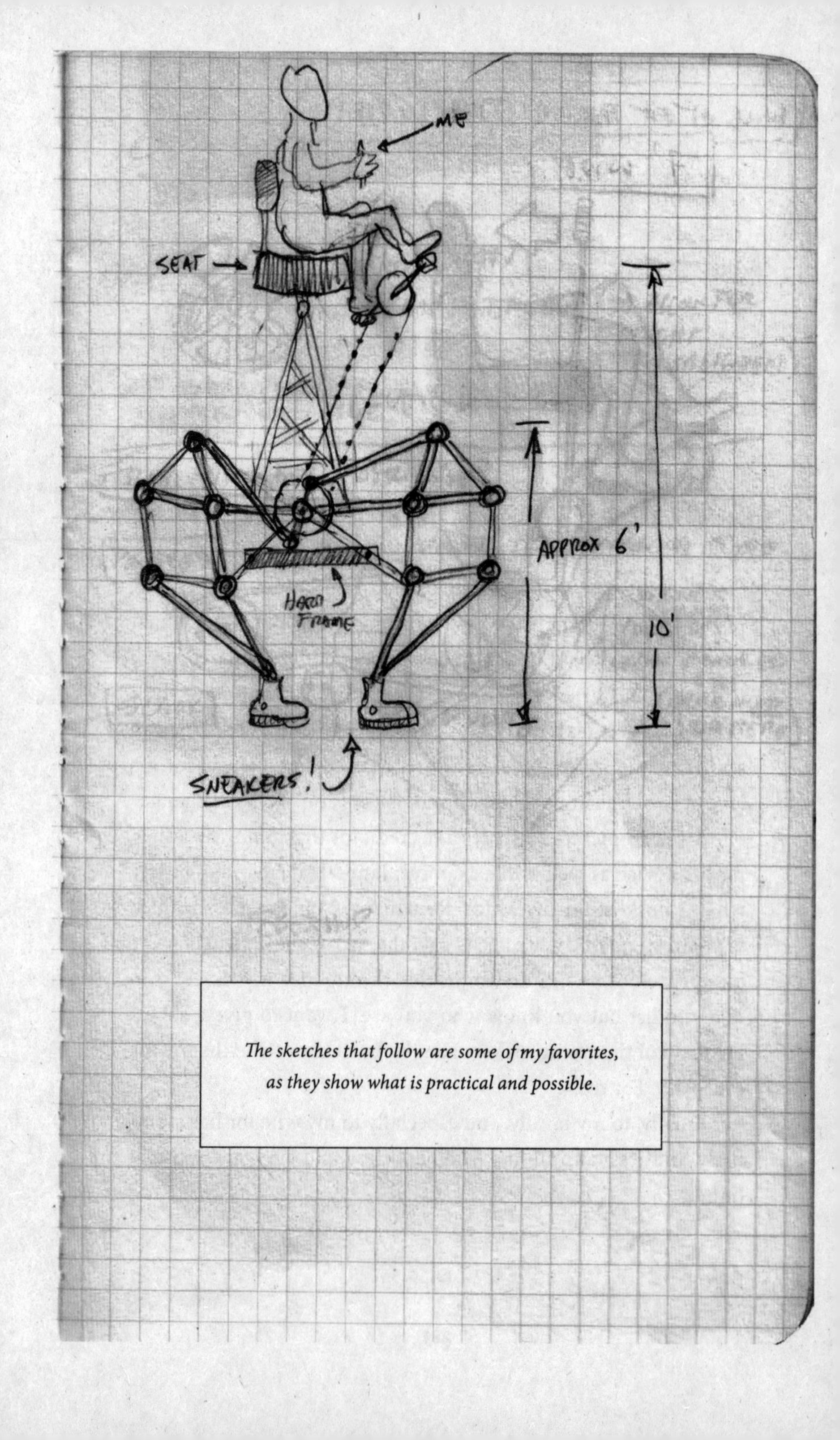

The sketches that follow are some of my favorites, as they show what is practical and possible.

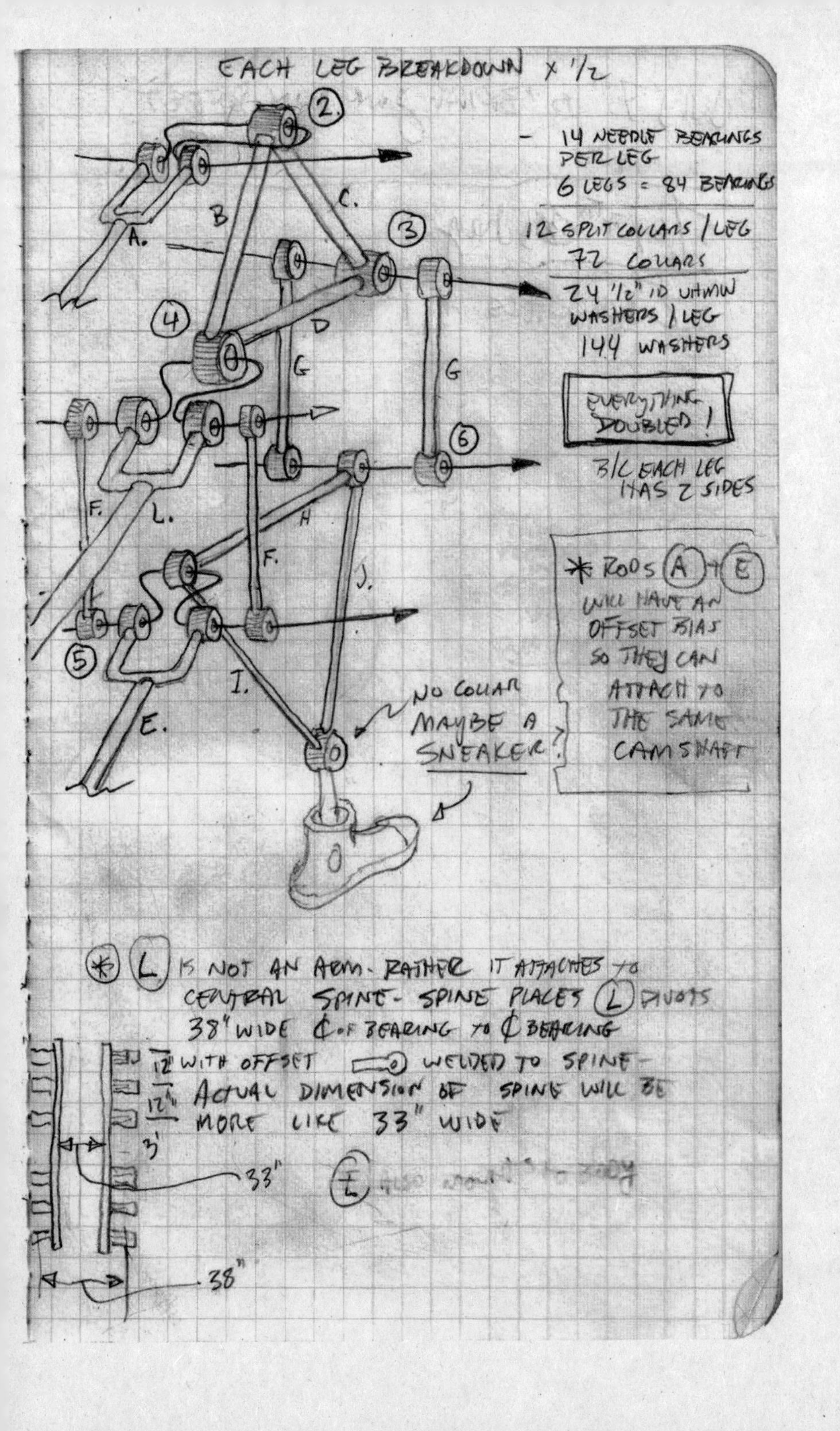
EACH LEG BREAKDOWN x 1/2
2
3
4
5
6
A.
B
C.
D
E.
F.
G
H
I.
J.
L.
- 14 NEEDLE BEARINGS PER LEG
6 LEGS = 84 BEARINGS
12 SPLIT COLLARS / LEG
72 COLLARS
24 1/2" ID UHMW WASHERS / LEG
144 WASHERS
EVERYTHING DOUBLED!
B/C EACH LEG HAS 2 SIDES
* RODS A + E WILL HAVE AN OFFSET BIAS SO THEY CAN ATTACH TO THE SAME CAMSHAFT
NO COLLAR MAYBE A SNEAKER?
* L IS NOT AN ARM - RATHER IT ATTACHES TO CENTRAL SPINE - SPINE PLACES L PIVOTS 38" WIDE ¢ OF BEARING TO ¢ BEARING
12" WITH OFFSET WELDED TO SPINE -
12" ACTUAL DIMENSION OF SPINE WILL BE MORE LIKE 33" WIDE
3'
33"
38"

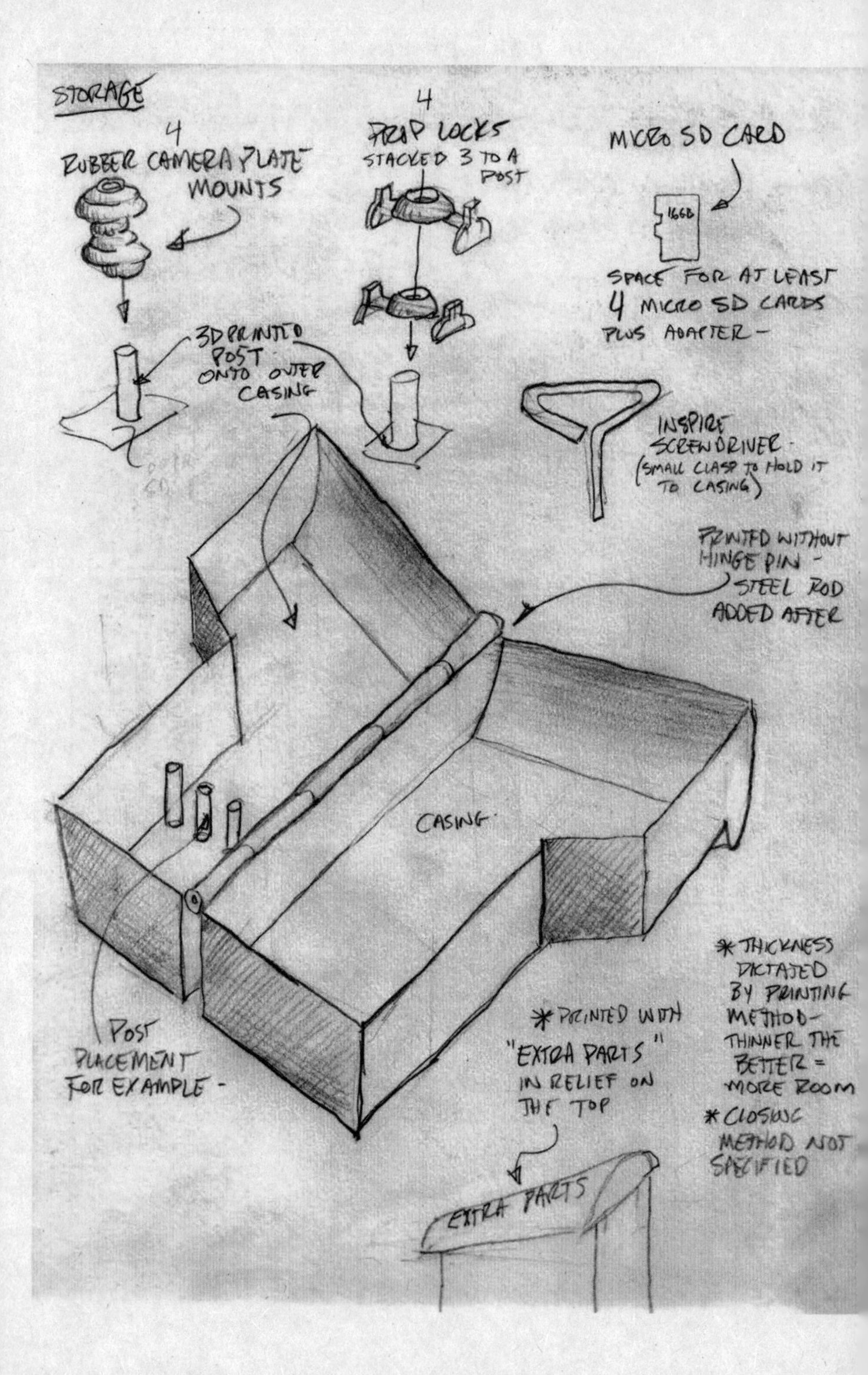
STORAGE
4
RUBBER CAMERA PLATE MOUNTS
4
PROP LOCKS
STACKED 3 TO A POST
MICRO SD CARD
16GB
SPACE FOR AT LEAST 4 MICRO SD CARDS PLUS ADAPTER -
3D PRINTED POST ONTO OUTER CASING
INSPIRE SCREWDRIVER
(SMALL CLASP TO HOLD IT TO CASING)
PRINTED WITHOUT HINGE PIN - STEEL ROD ADDED AFTER
CASING
POST PLACEMENT FOR EXAMPLE -
* PRINTED WITH "EXTRA PARTS" IN RELIEF ON THE TOP
* THICKNESS DICTATED BY PRINTING METHOD - THINNER THE BETTER = MORE ROOM
* CLOSING METHOD NOT SPECIFIED
EXTRA PARTS

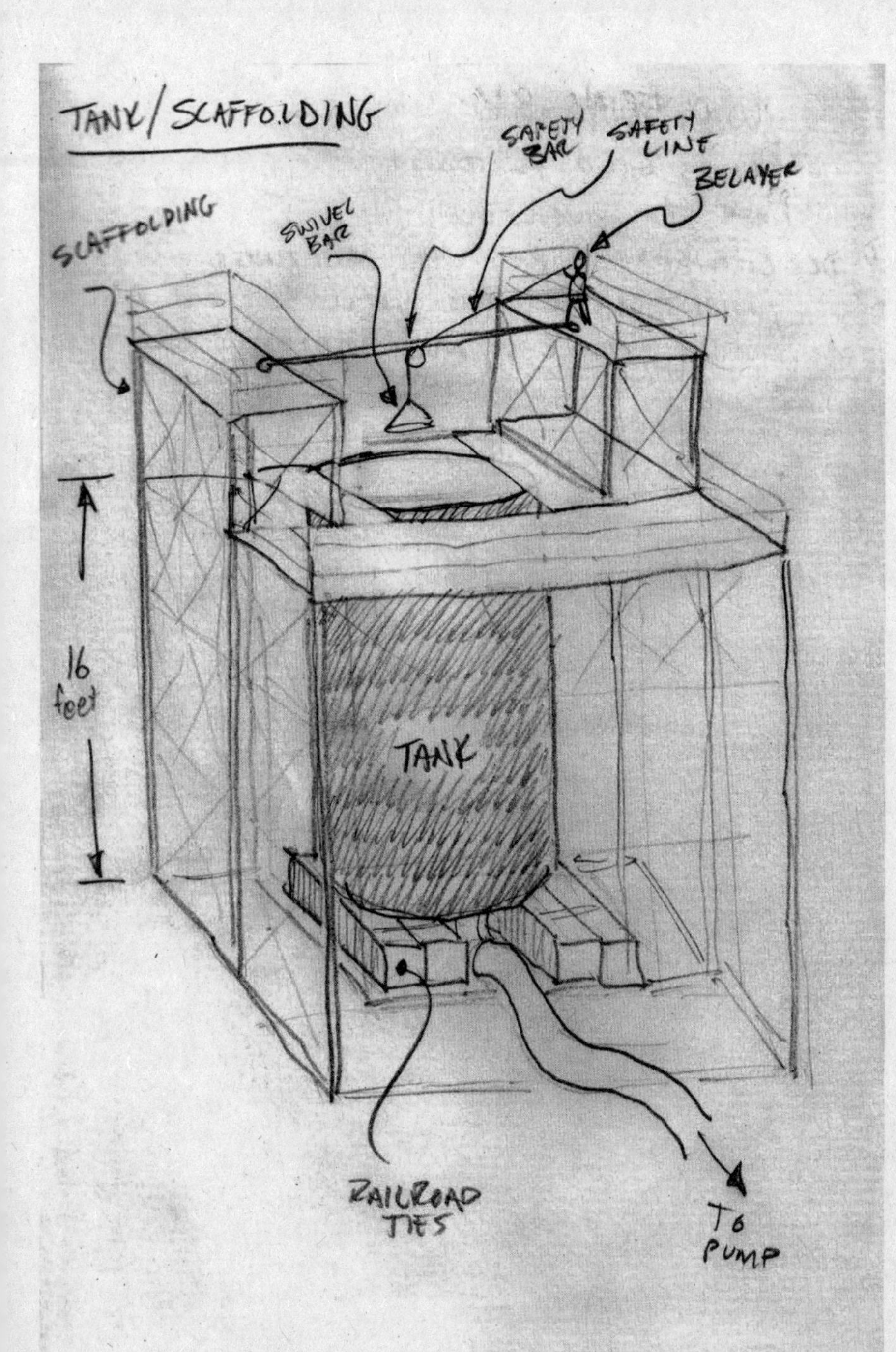

TANK/SCAFFOLDING
SAFETY BAR
SAFETY LINE
BELAYER
SCAFFOLDING
SWIVEL BAR
16 feet
TANK
RAILROAD TIES
TO PUMP

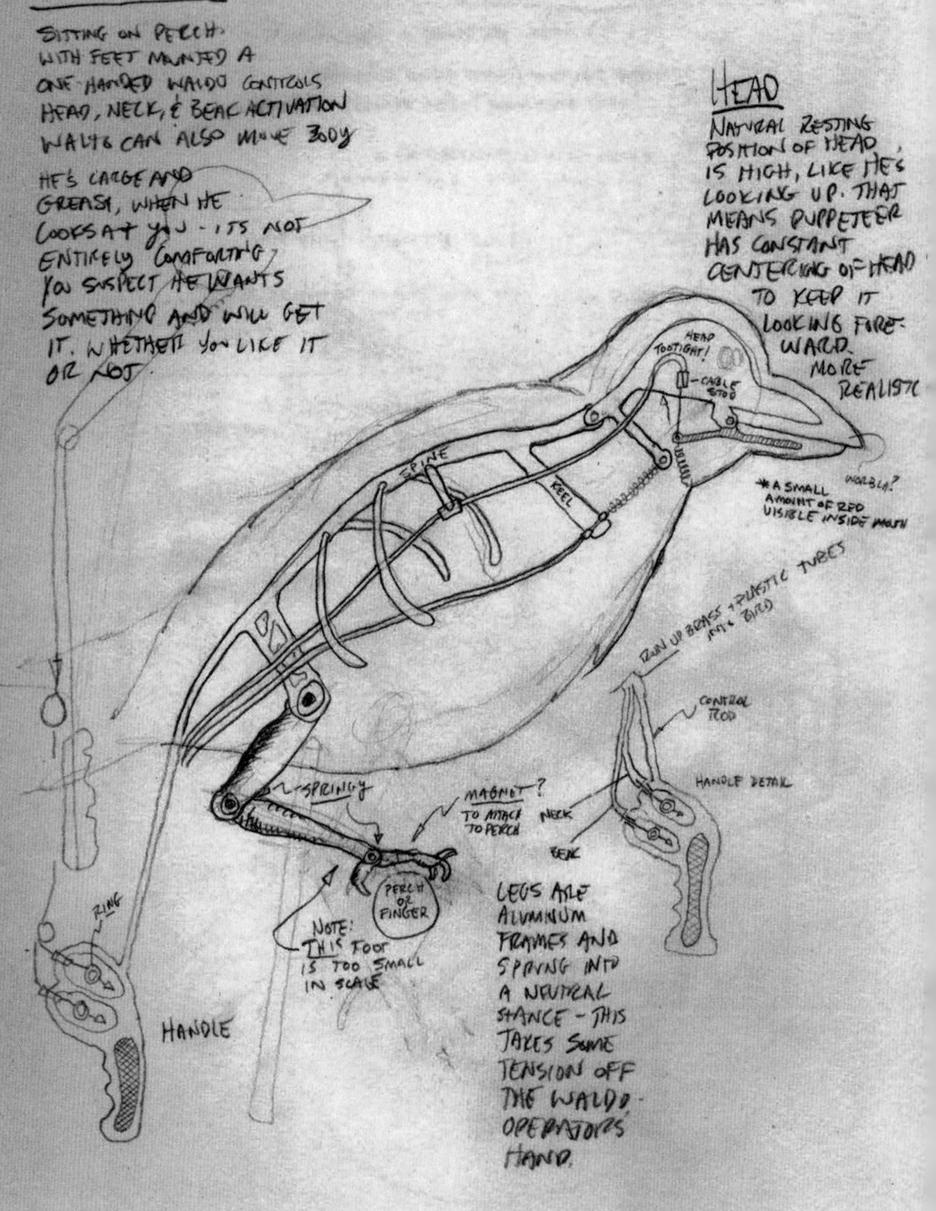
RAVEN PUPPET
SITTING ON PERCH.
WITH FEET MOUNTED A
ONE HANDED WALDO CONTROLS
HEAD, NECK, & BEAK ACTIVATION
WALDO CAN ALSO MOVE BODY
HE'S LARGE AND
GREASY, WHEN HE
LOOKS AT YOU - ITS NOT
ENTIRELY COMFORTING
YOU SUSPECT HE WANTS
SOMETHING AND WILL GET
IT. WHETHER YOU LIKE IT
OR NOT.
HEAD
NATURAL RESTING
POSITION OF HEAD
IS HIGH, LIKE HE'S
LOOKING UP. THAT
MEANS PUPPETEER
HAS CONSTANT
CENTERING OF HEAD
TO KEEP IT
LOOKING FORE-
WARD.
MORE
REALISTIC
HEAD TOO TIGHT!
CABLE STOP
SPINE
REEL
WOBBLY?
*A SMALL
AMOUNT OF RED
VISIBLE INSIDE MOUTH
RUN UP BRASS + PLASTIC TUBES
INTO BIRD
CONTROL ROD
HANDLE DETAIL
NECK
BEAK
SPRING
MAGNET?
TO ATTACH
TO PERCH
PERCH
OR
FINGER
NOTE:
THIS FOOT
IS TOO SMALL
IN SCALE
LEGS ARE
ALUMINUM
FRAMES AND
SPRING INTO
A NEUTRAL
STANCE - THIS
TAKES SOME
TENSION OFF
THE WALDO
OPERATORS'
HAND.
RING
HANDLE

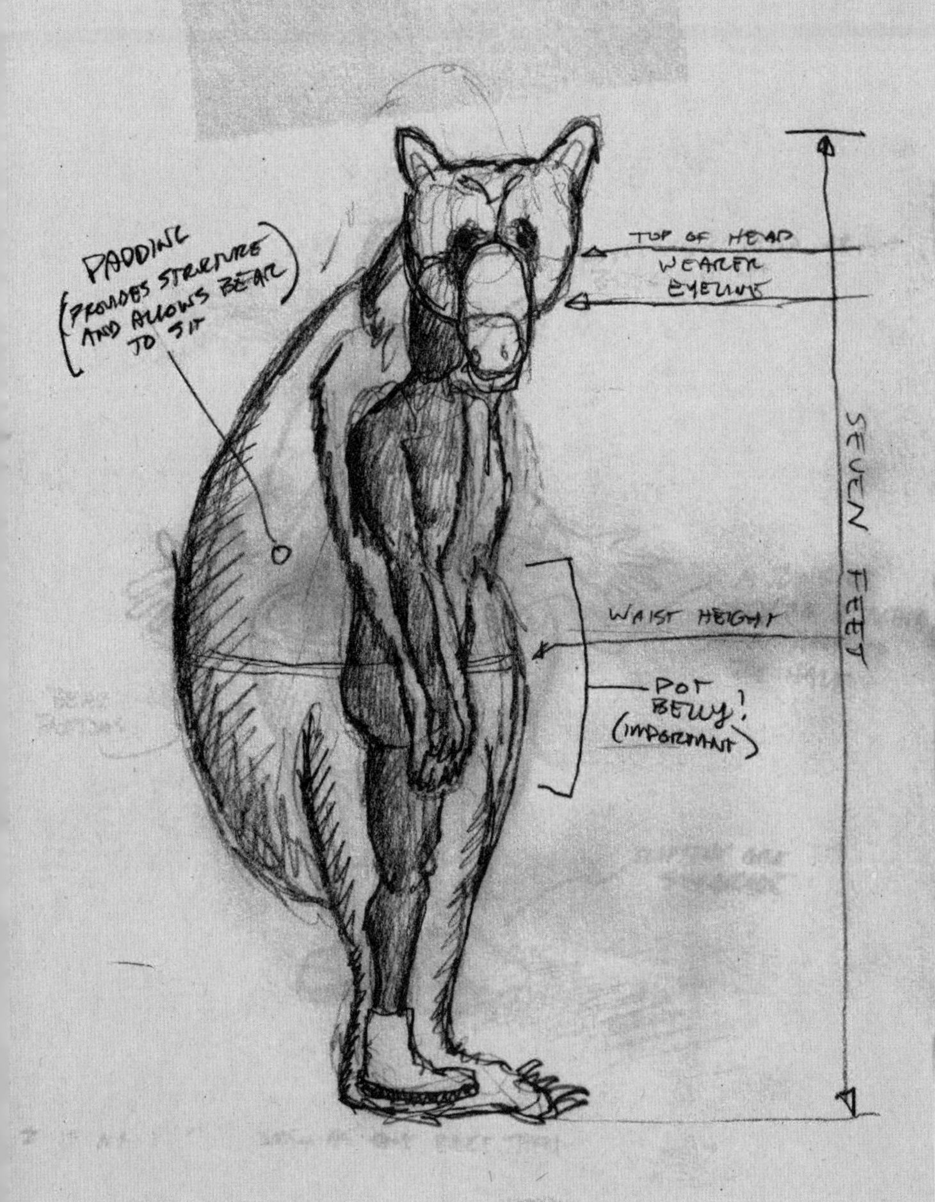
PADDING
(PROVIDES STRUCTURE AND ALLOWS BEAR TO SIT)
TOP OF HEAD
WEARER EYELINE
SEVEN FEET
WAIST HEIGHT
POT BELLY!
(IMPORTANT)

3.

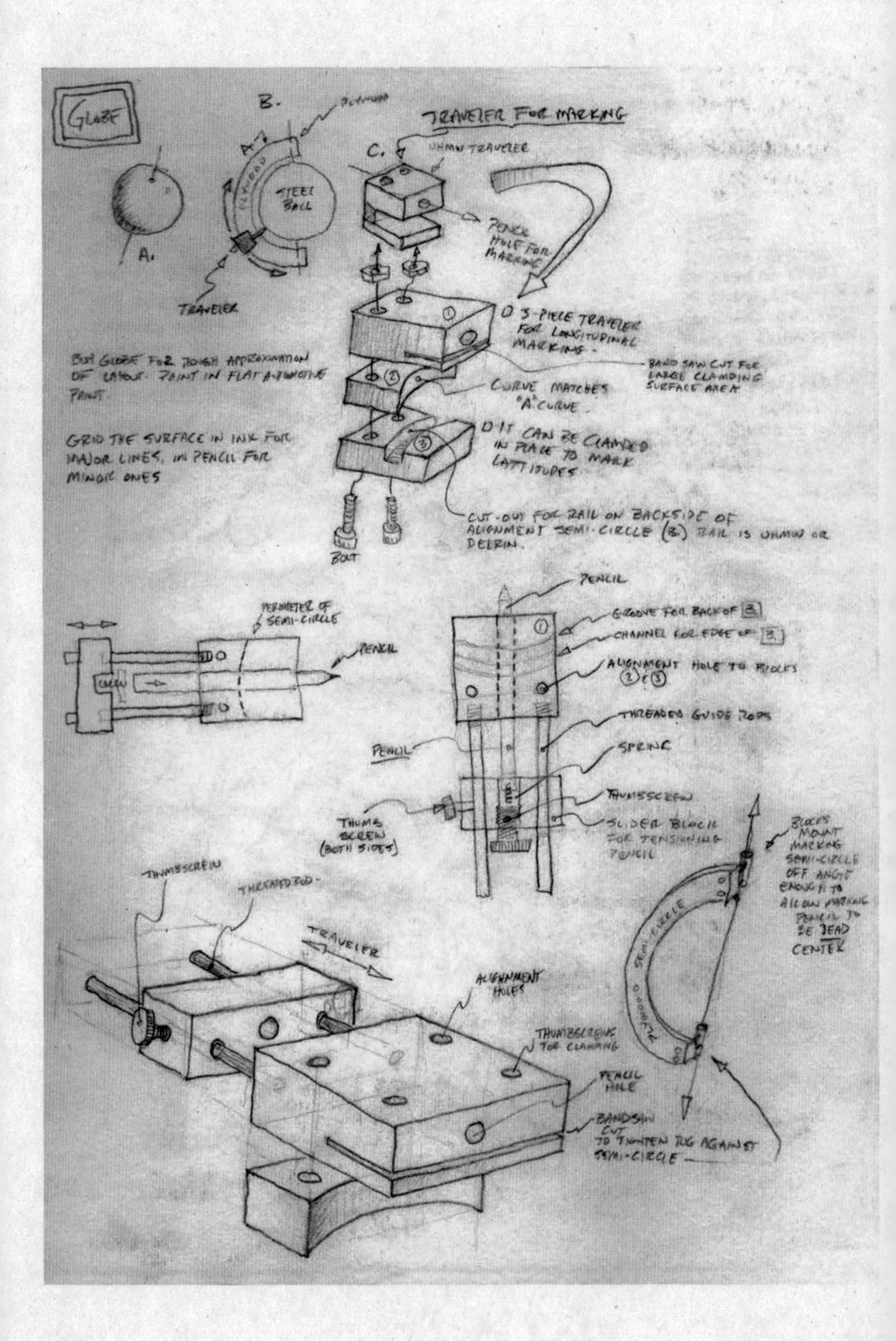
GLOBE
B.
PLYWOOD
TRAVELER FOR MARKING
C.
UHMW TRAVELER
STEEL BALL
PLYWOOD
A.
TRAVELER
PENCIL HOLE FOR MARKING
3-PIECE TRAVELER FOR LONGITUDINAL MARKING.
BAND SAW CUT FOR LARGE CLAMPING SURFACE AREA
CURVE MATCHES "A" CURVE
IT CAN BE CLAMPED IN PLACE TO MARK LATITUDES
CUT-OUT FOR RAIL ON BACKSIDE OF ALIGNMENT SEMI-CIRCLE (B.) RAIL IS UHMW OR DELRIN.
BOLT
GRID THE SURFACE IN INK FOR MAJOR LINES, IN PENCIL FOR MINOR ONES
PERIMETER OF SEMI-CIRCLE
PENCIL
PENCIL
GROOVE FOR BACK OF B.
CHANNEL FOR EDGE OF B.
ALIGNMENT HOLE TO BLOCKS 2 & 3
THREADED GUIDE RODS
PENCIL
SPRING
THUMBSCREW
THUMB SCREW (BOTH SIDES)
SLIDER BLOCK FOR TENSIONING PENCIL
THUMBSCREW
THREADED ROD
TRAVELER
ALIGNMENT HOLES
THUMBSCREWS FOR CLAMPING
PENCIL HOLE
BANDSAW CUT TO TIGHTEN RIG AGAINST SEMI-CIRCLE
BLOCKS MOUNT MARKING SEMI-CIRCLE OFF ANGLE ENOUGH TO ALLOW MARKING PENCIL TO BE DEAD CENTER
SEMI-CIRCLE
PLYWOOD

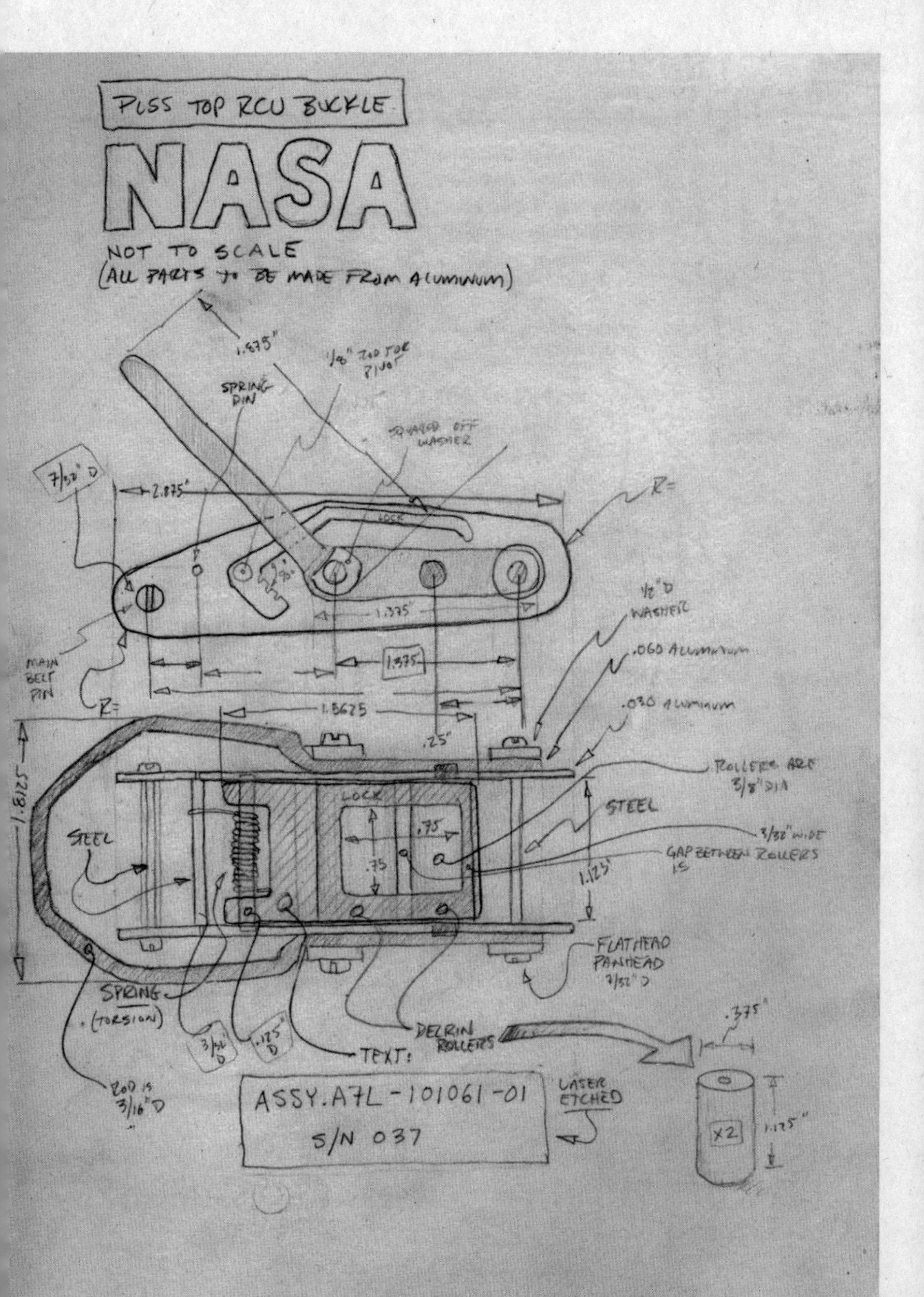

PLSS TOP RCU BUCKLE
NASA
NOT TO SCALE
(ALL PARTS TO BE MADE FROM ALUMINUM)
1.875"
1/8" ROD FOR PIVOT
SPRING PIN
SQUARED OFF WASHER
7/32" D
2.875"
R=
LOCK
1.375"
1/2" D WASHER
.060 ALUMINUM
MAIN BELT PIN
1.375
R=
1.8625
.030 ALUMINUM
.25"
1.8125
ROLLERS ARE 3/8" DIA
STEEL
LOCK
.75
.75
STEEL
3/32" WIDE GAP BETWEEN ROLLERS IS
1.125"
FLATHEAD PANHEAD 7/32" D
SPRING (TORSION)
3/16" D
.125" D
DELRIN ROLLERS
TEXT:
.375"
ROD IS 3/16" D
ASSY. A7L-101061-01
S/N 037
LASER ETCHED
X2
1.125"

INDEX

Page numbers in *italics* refer to images.

ABOUT THE AUTHOR

ADAM SAVAGE is a maker, designer, television host, producer, husband, and father. He was the cohost of all 278 hours of *Myth-Busters* on the Discovery Channel for fourteen years and host of its 2019 spinoff *MythBusters Jr.*, as well as several other TV shows. He also makes stuff and tells his stories on his website Tested.com. He lives in San Francisco with his wife, twin boys, and two amazing dogs. *Every Tool's a Hammer* is his first book.